Animal Cell Biotechnology

Animal Cell Biotechnology

Harry Johnson

STATES
ACADEMIC PRESS
www.statesacademicpress.com

Animal Cell Biotechnology
Harry Johnson
ISBN: 978-1-63989-048-4 (Hardback)

STATES
ACADEMIC PRESS

Published by States Academic Press,
109 South 5th Street,
Brooklyn, NY 11249, USA

Cataloging-in-Publication Data

Animal cell biotechnology / Harry Johnson.
p. cm.
Includes bibliographical references and index.
ISBN 978-1-63989-048-4
1. Animal cell biotechnology. 2. Cell culture. I. Johnson, Harry.
TP248.27.A53 A55 2022
660.6--dc23

For more information regarding States Academic Press and its products, please visit the publisher's website www.statesacademicpress.com

Table of Contents

Preface

The branch of biology which makes use of living organisms to create new products is termed as biotechnology. The application of biotechnology for studying animal cells and cell culture is termed as animal cell biotechnology. Some of the key areas of study under this field are sequencing of animal genomes, recombinant DNA technology, identification of genes, regulation of genes, molecular markers and diagnostics. The animals which are genetically altered to have specific characteristics are known as transgenic animals. They are produced using a variety of techniques such as nuclear transfer, DNA microinjection, sperm mediated gene transfer and artificial chromosomes. Animal cell biotechnology finds application across various fields such as food production, organ transplant, pharmaceuticals, animal forensics and animal breeding. Animal cell biotechnology is an upcoming field of science that has undergone rapid development over the past few decades. Most of the topics introduced herein cover new techniques and the applications of animal cell biotechnology. Those in search of information to further their knowledge will be greatly assisted by this book.

To facilitate a deeper understanding of the contents of this book a short introduction of every chapter is written below:

Chapter 1- The area of biotechnology which makes use of animal cells that are propagated in vitro, for the manufacture of biological molecules, and for the discovery and testing of medicine, is known as animal cell biotechnology. Some of the topics which fall under this field of study are animal cell, animal biotechnology, transcriptomics and proteomics, etc. This is an introductory chapter which will briefly introduce all these significant aspects of animal cell biotechnology.

Chapter 2- The process of growing animal cells in vitro on a favorable artificial medium and in a favorable environment is known as animal cell culture. There are various types of animal cell culture such as primary and secondary cell culture, and subculture of cell. This chapter explains all these types of animal cell cultures and various other aspects related to it such as organ culture, cell cryopreservation, etc.

Chapter 3- The methods used to incorporate foreign genes into the genome of an animal so that these genes are inherited and expressed in the offspring fall under transgenic animal technology. Transgenic animals are those animals whose genome has been modified to have desired traits. This chapter elaborates methods of creation of transgenic animals, and applications and ethical concerns of transgenic technology.

Chapter 4- Stem cells are the cells which have the potential to differentiate into several specific cell types. Stem cell technology involves the use of these stem cells for the treatment of a wide variety of diseases. Stem cells are broadly divided into totipotent stem cells, pluripotent stem cells and mesenchymal stem cells. The topics elaborated in this chapter will help in gaining a better perspective about stem cell technology and stem cell therapy.

Chapter 5- Genetic engineering is a set of technologies used to modify the genome of organisms. It includes the transfer of genes from one organism to another within or across species. This chapter discusses in detail the major concepts of genetic engineering such as gene expression, gene cloning, recombinant proteins, and cell and tissue engineering.

I would like to share the credit of this book with my editorial team who worked tirelessly on this book. I owe the completion of this book to the never-ending support of my family, who supported me throughout the project.

Harry Johnson

1

Introduction to Animal Cell Biotechnology

The area of biotechnology which makes use of animal cells that are propagated in vitro, for the manufacture of biological molecules, and for the discovery and testing of medicine, is known as animal cell biotechnology. Some of the topics which fall under this field of study are animal cell, animal biotechnology, transcriptomics and proteomics, etc. This is an introductory chapter which will briefly introduce all these significant aspects of animal cell biotechnology.

Animal Cell

An animal cell is a unique factory, where thousands of genes are encoded and transcribed; products are translated, and finally processed to biologically active molecules. It is therefore, important to understand inside of a cell, how cellular functions are coordinated, limitation of cells, reasons for proliferation and cellular death.

Animal cells are the basic unit of life in organisms of the kingdom Animalia. They are eukaryotic cells, meaning that they have a true nucleus and specialized structures called organelles that carry out different functions. Animal cells do not have plant-specific organelles like cell walls, which support the plant cell, or chloroplasts, the organelle that carries out photosynthesis.

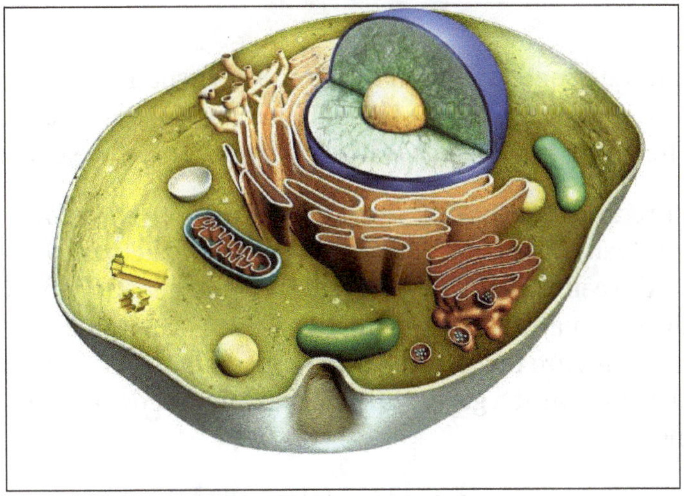

Figure: 3D model of a typical animal cell.

Animal Cell Structure

Animal cells have a variety of different organelles that work together to allow the cell to perform its functions. Each cell can be thought of as a large factory with many departments, like manufacturing, packaging, shipping, and accounting. Different organelles represent each of these departments.

There are lots of different animal cells that each carry out specialized functions. Therefore, not every animal cell has all types of organelles, but in general, animal cells do contain most (if not all) of the following organelles. Additionally, some organelles will be highly abundant in certain cells and not others.

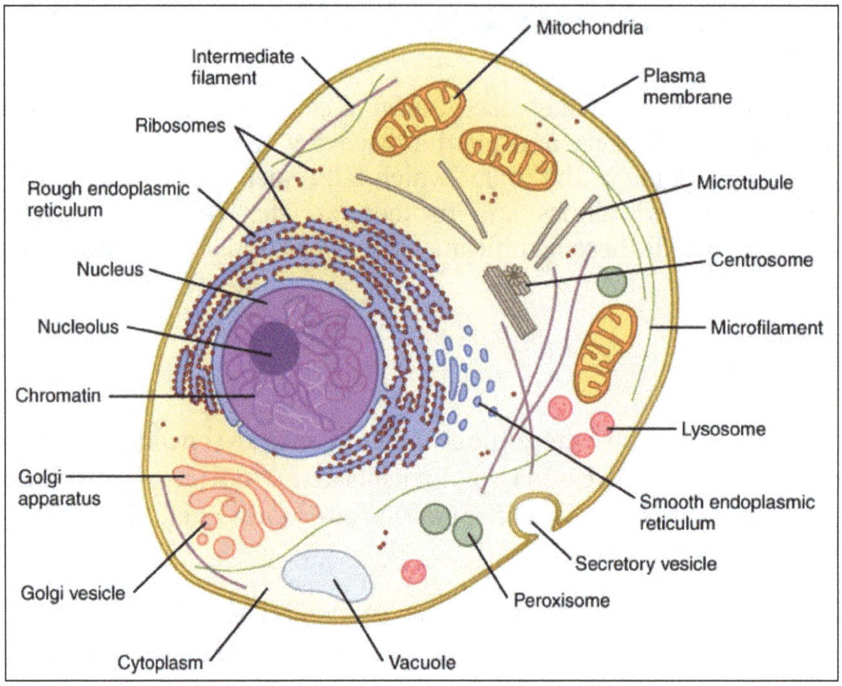

Figure: Labeled diagram of a typical animal cell.

Nucleus

The nucleus contains all the genetic material in a cell. This genetic information is called deoxyribonucleic acid (DNA). DNA contains all the instructions for making proteins, which control all of the body's activities. Therefore, the nucleus is like the manager's office of the cell.

DNA is an extremely precious and tightly regulated molecule. Therefore, it does not just exist naked in the nucleus! Instead, DNA is tightly wound around structural proteins called histones to form chromatin. When the cell is ready to divide to pass the genetic information on to new cells (the daughter cells), the chromatin forms highly condensed structures called chromosomes. The nucleus regulates which genes are turned 'on' in the cell, and at what time. This controls the cell's activity. The genes that are active at a given time will be different depending on the type of cell and the function it performs.

The nucleus is surrounded by a nuclear envelope (also called the nuclear membrane), which separates it from the rest of the cell. The nuclear envelope also contains pores that permit the entry and exit of some molecules.

As well as all the genetic material, there is also a sub-section of the nucleus called the nucleolus, which looks like a nucleus within the nucleus. The nucleolus is the site of ribosome synthesis. The nucleus is surrounded by a nuclear envelope (also called nuclear membrane), which separates it from the rest of the cell.

The nucleus also regulates the growth and division of the cell. When the cell is preparing to divide during mitosis, the chromosomes in the nucleus duplicate and separate, and two daughter cells form. Organelles called centrosomes help to organize the DNA during cell division.

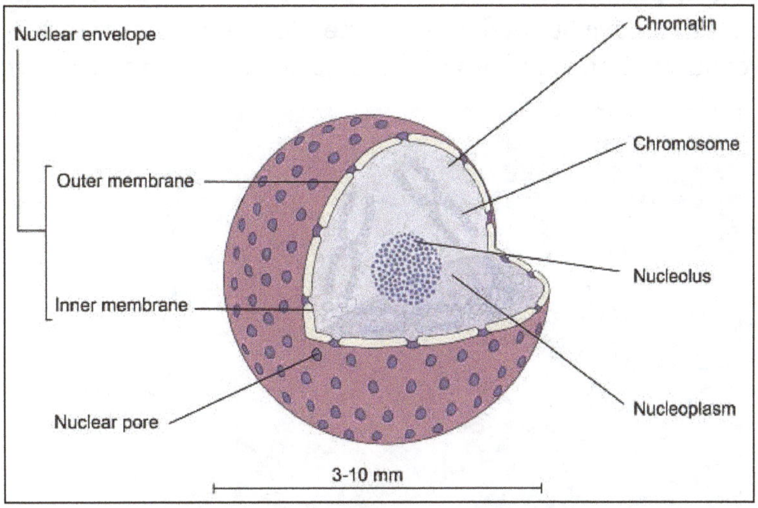

Figure: The nucleus contains DNA in the form of chromatin. Chromatin can be further compacted to form chromosomes. The nucleus is surrounded by a double envelope that contains pores to allow certain materials to pass in and out. The nucleus also contains a region called the nucleolus.

Ribosomes

Ribosomes are organelles found in both prokaryotic and eukaryotic cells. They are like mini machines that synthesize all the proteins in the cell. In any single animal cell, there can be as many as 10 million ribosomes! The ribosomes form the manufacturing department of the cell.

In the nucleus, a sequence of DNA that codes for a specific protein is copied onto an intermediate molecule called messenger RNA (mRNA). The mRNA molecule carries this information to the ribosome, and its sequence determines the order of amino acids in a polypeptide chain. The ribosome synthesizes this polypeptide chain, which eventually folds to become a protein. In animal cells, ribosomes can be found freely in a cell's cytoplasm or attached to the endoplasmic reticulum.

Endoplasmic Reticulum

The endoplasmic reticulum (ER) is a network of flattened, membrane-bound sacs that are involved in the production, processing, and transport of proteins that have been synthesized by ribosomes. The endoplasmic reticulum is like the assembly line of the cell, where the products produced by the ribosomes are processed and assembled.

There are two kinds of endoplasmic reticulum: smooth and rough. The rough ER has ribosomes attached to the surface of the sacs. Smooth ER does not have ribosomes attached and has functions in storage, synthesizing lipids, removing toxic substances.

Golgi Apparatus

The Golgi apparatus, also called the Golgi complex or Golgi body, receives proteins from the ER and folds, sorts, and packages these proteins into vesicles. The Golgi apparatus is like the shipping department of the cell, as it packages proteins up for delivery to their destinations.

Like the ER, the Golgi apparatus also consists of a series of membrane-bound sacs. These sacs originate from vesicles that have budded off from the ER. Unlike the system of membranes in the ER, which are interconnected, the pouches of the Golgi apparatus are discontinuous.

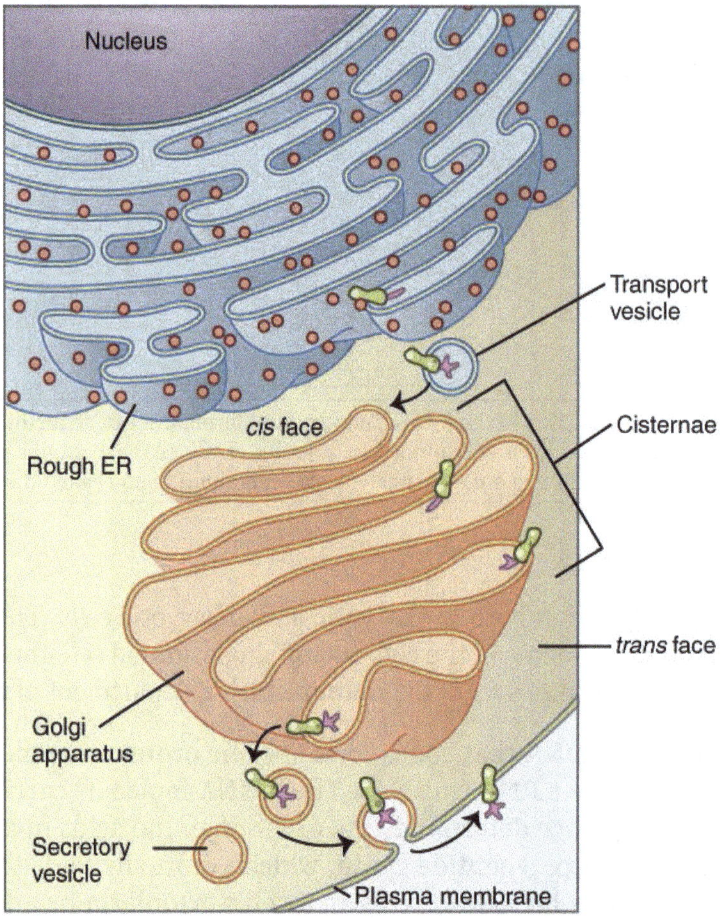

Figure: Comparison of the functions of the endoplasmic reticulum and the Golgi apparatus.

Lysosomes

Lysosomes are a type of vesicle. Vesicles are spheres surrounded by a membrane that excludes their contents from the rest of the cytoplasm. Vesicles are used extensively within the cell for metabolism and transport of large molecules that cannot cross membrane unaided. Lysosomes are specialized vesicles that contain digestive enzymes. These enzymes can break down large molecules like organelles, carbohydrates, lipids, and proteins into smaller units so that the cell can reuse them. Therefore, they are like the waste disposal/recycling department of the cell.

Mitochondria

Mitochondria are the energy-producing organelles, commonly known as "the powerhouse of the cell." The process of cellular respiration occurs in the mitochondria. During this process, sugars and fats are broken down through a series of chemical reactions, releasing energy in the form of adenosine triphosphate (ATP). ATP is like the energy currency of the cell. Think of each molecule like a rechargeable battery that can be used to power various cellular processes.

Cytoplasm

The cytosol is the gel-like liquid contained within cells. The cytosol and all the organelles within it – except for the nucleus – are collectively referred to as the cell's cytoplasm. This cytosol consists primarily of water, but also contains ions, proteins, and small molecules. The pH is generally neutral, around 7.

Cytoskeleton

The cytoskeleton is a network of filaments and tubules found throughout the cytoplasm of the cell. It has many functions: it gives the cell shape, provides strength, stabilizes tissues, anchors organelles within the cell, and has a role in cell signaling. It also provides mechanical support to allow cells to move and divide. There are three types of cytoskeletal filaments: microfilaments, microtubules, and intermediate filaments.

Cell Membrane

The cell membrane surrounds the entire cell and separates its components from the outer environment. The cell membrane is a double layer made up of phospholipids (called the phospholipid bilayer). Phospholipids are molecules with a phosphate group head attached to glycerol and two fatty acid tails. They spontaneously form double membranes in water due to the hydrophilic properties of the head and hydrophobic properties of the tails.

The cell membrane is selectively permeable, meaning it only allows certain molecules to enter and exit. Oxygen and carbon dioxide pass through easily, while larger or charged molecules must go through special channels, bind to receptors, or be engulfed.

Animal Biotechnology

Animal biotechnology is defined as the application of biotechnology to the processing or production of materials by animals (or aquatic species) to provide goods and services. Scientists are using biotechnology to address some of the major global problems of modern times including the food crisis, health and disease, and conservation, through advances in the following areas:

- Agriculture:
    - Food production
    - Animal breeding
    - Animal health and nutrition

- Medicine:
    - Transplant organs
    - Pharmaceuticals
    - Research into human disease
- Industrial applications
- Conservation and ecology
- Animal forensics

Various technologies are employed in each of these areas and recent progresses in such technologies have facilitated much development in animal biotechnology. These include sequencing and analysis of animal genomes, identification of genes and their variations and an understanding of how these genes are regulated, advances in molecular markers and diagnostics.

Recombinant DNA technology has produced transgenic animals (i.e. animals that contain foreign genes for a specific purpose). The first transgenic animal produced in 1982 was a mouse. Most transgenic animals that have been produced since are mice, used in medical research.

As the technology has developed, other transgenic animals have been produced for economic purposes. Cloning technologies have been developed to produce offspring genetically identical to the parent.

Transgenesis

A transgenic animal is one which has been genetically altered to have specific characteristics it otherwise would not have. In animals, transgenesis either means transferring DNA into the animal or altering DNA already in the animal.

There are various techniques that can be used to create transgenic animals. Nuclear transfer involves the transfer of a nucleus into a single egg whose nucleus has been removed. Somatic Cell nuclear transfer is a method used to create clones of animals. Transgenesis differs from cloning in that the nuclues injected into the cell would contain genetically modified DNA. In cloning, the nucleus would be unmodified from the parent animal. DNA microinjection of embryos has been the traditional method for generating transgenic livestock. This involves the physical injection of a DNA solution into the pronuclei of a zygote. The DNA in solution then integrates randomly into the cellular DNA, making the process a very inefficient process with a success rate of less than 4%. Sperm-mediated gene transfer uses sperm to carry transgenic DNA to fertilise an oocyte. The sperm is pre-incubated with DNA which is randomly picked up by the sperm. This process is also inefficient; however improved uptake of DNA by sperm improve the success rate of creating transgenic animals. Homologous recombination is a method generally used to delete a gene. Retrovirus-mediated gene transfer uses the natural gene delivery system of the retrovirus to transfer DNA as it infects the host cell, which then becomes integrated in the host cell genome. Artificial chromosomes are large self-replicating constructs which allow for the transfer of large DNA constructs and ensure better control of transgene expression. The disadvantage is the difficulty in handling such large fragments of DNA.

In addition to the inefficiency in the success of transfer of foreign DNA and creating a transgenic animal, another limitation in the technology is the stability of the expression of the introduced gene. Many of the genes introduced are expressed at low levels or non-specifically. Also, the effects have been reported to decrease over time in the individual or in subsequent generations.

Other concerns regarding the technology are the environmental stability/safety. This refers to the potential for the gene to spread to the wild natural populations should a transgenic animal escape or be released, as well as the potential to spread to other animals including humans through consumption of its meat or milk. Another concern is the potential for novel disease being intiated through the process of transgenesis. Also, various ethical concerns have been raised. All these areas of concern are the target of research into the process of transgenesis.

Agriculture

Food Production

Due to the population explosion, we exist in a time of global food crisis, where for the past decade the world has been consuming more food than it has been producing. New technologies are being developed to increase food production, to produce more food of higher quality in a shorter period of time. This includes animal biotechnology being applied to livestock to produce animals with a greater size, improved lean muscle mass, increased growth rate, greater milk production, and production of milk with enhanced qualities such as higher protein content or lower fat content. Another example would be chickens with higher quality eggs. Traditional breeding to select and propagate animals with such desired characteristics is a slow and arduous process. Biotechnology has the potential to short-cut traditional/ conventional selection/breeding methods to create animals with the desired traits. An example is seen in transgenic fish created for use in aquaculture fish farming. A growth hormone transgene has been introduced into many species of fish, with many showing increased growth rates. Coho salmon with a growth hormone transgene showed an increase in growth rate so that fish reached market size in two years in comparison to normal controls which took three years. The fish did not grow beyond the normal adult size finally. A company called Aqua Bounty is engineering salmon, trout and tilapia to grow faster than traditional fish.

In the United States, most transgenic animals are regulated by one or more of the FDA (Food and Drug Administration) centres. At this stage no transgenic animals have been approved for use as human food.

Animal Breeding

Many animal genomes have now been completely sequenced and a number of gene variants have been described and can be identified. Biotechnology can be used to identify and select animals with mulitple desirable characteristics, particularly traits associated with economic importance, and the best combination of gene variants for breeding.

The University of Pretoria, Department of Animal and Wildlife Sciences is researching the genetics of Angora goats, particularly to identify a Quantitative. Trait Loci (QTL) that is associated with the trait of fine mohair. Being able to identify individuals with a particular makeup with regard to the QTL will improve the efficiency of selection for breeding.

Cloning is becoming a viable method of reproducing desired animals of many species. Animals that have been cloned include sheep, pigs, cows, goats, horses, mules, cats, rat's amd mice. Not only does this have potential significance in maintaining the desired characteristics in livestock for food production (which is debatable of course!), but also has potential significance in conservation of endangered species.

Animal Health

Although creating disease resistance in animals is not a practical reality as yet, as a limited number of genes are known to be responsible for disease resistance, biotechnology is making advances in reduction of diseases. This also includes development of vaccines against significant animal diseases.

The Onderstepoort Veterinary Institute (OVI) near Pretoria has made significant advances in the development of vaccines against key endemic animal diseases. Research at the OVI has led to the development of unique vaccines against African horsesickness virus in horses, bluetongue virus in sheep, botulism, anthrax, lumpy skin disease and more. At the University of Pretoria the Department of Genetics has a key research programme into the molecular biology of orbiviruses, largely African horsesickness virus. The aim is to understand important aspects of the virus and its interaction with different cells, its replication process and its virulence, in order to provide insight into the development of new and more successful vaccines.

Medicine

Transplant Organs

With a shortage of suitable organ donors for failing or diseased organs, science and medicine is looking to other species to make up the shortage. "Xenotransplantation" refers to the process of transplanting organs, tissues or cells between different species. The most promising animal for xenotransplantation to humans is the pig, with organs of similar size to those of humans. However, without genetic manipulation and biotechnology, the pig would not be immunologically incompatible as a donor and organs would be rejected.

Biotechnology and gene knockout technology (in which a gene is deleted or inactivated) is developing pigs with immunologically compatible organs. One of the major problems in pigs being immunologically incompatible is caused by a type of sugar molecule called Gal-alpha(1,3)-Gal which is attached to protein molecules on the surface of pig cells but not on human cells. In humans it is recognised as foreign, leading to a rapid rejection of the tissue called hyperacute rejection. In transgenic pigs created for the purpose of organ donation, the gene for the sugar molecule has been "knocked out", so that the sugar molecule is not expressed on the cells of the pig.

There are various ethical issues when looking at pigs as a source of donor organs and these issues will need to be addressed as the possibility of using pig organs in humans nears reality. Public education is essential, informed consent and the selection of human donors must be carefully considered, and religious beliefs must not be ignored. Safety is a critical issue. Organs and cells may be rejected, the survival and longevity of xenotransplanted organs are unknown, and additionally, the possibility exists of novel infections being transmitted from the donor animal to the human recipient. There will be many regulatory stages to pass before this technology becomes a real possibility.

Pharmaceuticals

Transgenic poultry, pigs, goats and cattle have been produced that produce large quantities of human protein in eggs, milk, blood or urine with the intention to use these products as pharmaceuticals such as enzymes, albumin, clotting factors and antibodies. In a sense, these transgenic animals can be seen as living bioreactors for the production of products of value.

The USA's Food and Drug Administration (FDA) has recently approved the first pharmaceutical produced in genetically engineered animals, called ATryn, manufactured by GTC Biotherapeutics Inc. ATryn is produced in the milk of transgenic goats which have been genetically engineered to contain the gene to produce human antithrombin in its milk. Antithrombin (AT) is a protein that helps to prevent blood clotting in the veins and arteries of healthy individuals. Patients with a rare disease called hereditary AT deficiency lack normal AT. In these patients, ATryn acts as an anti-coagulant. Because this disease only occurs in a very small proportion of the population (approximately 1 in 5000 in the USA), the FDA had given ATryn an orphan drug designation, referring to a drug developed specifically to treat a rare medical condition.

ATryn was approved in 2006 by the European Medicines Agency for preventing clotting conditions in surgery in patients with hereditary AT deficiency. ATryn was approved by two FDA centres: The Center for Biologics Evaluation and Research (CBER) gave approval based on safety and efficacy, and the Center for Veterinary Medicine approved the rDNA construct in the goats, stating that seven generations of goats did not suffer adverse effects from the genetic modification.

Since the approval of ATryn, new guidelines have been released for the regulation of genetically engineered animals under the new animal drugs provisions of the Federal Food, Drug and Cosmetic Act (FFDCA), entitled "The Regulation of the Genetically Engineered Animals Containing Heritable rDNA constructs". This provides producers of GE animal's recommendations to help them comply with regulations and laws.

Other examples of therapeutic proteins from genetically engineered animals include the production of the protein 1-antitrypsin in sheep's milk. This protein is used to treat hereditary emphysema, cystic fibrosis, and chronic obstructive pulmonary disease which affect over 200 000 people in the US and Europe. This product is in clinical trials in Europe. Another example is human lysozyme produced in the milk of genetically engineered cows. Lysozyme is an antimicrobial protein and so can increase the shelf life of the milk and offer health protection to infants.

Research into Human Disease

Transgenic animals are useful as disease models. Animals are genetically manipulated using biotechnology to mimic symptoms of disease. In this way they can be used to study therapies and treatments. In particular, developments with transgenesis in mice will contribute to the further reduction in the use of "higher" animals, such as dogs and non-human primates, in biomedical research.

Approximately 80% of mouse genes have a single identifiable orthologue in humans, in other words genes of similar function that originated from a common ancestor. Mice have a short reproduction cycle and their embryos are easily manipulated. This makes them useful to study disease. Over 95% of transgenic animals used in biomedical research are mice. "Knockout" mice can be generated which carry specific mutations in a particular gene of interest. The physical result of that mutation can then be studied. This provides a direct way of studying the function of a gene.

A leading pharmaceutical company, GlaxoSmithKline will be using transgenic mice to carry out some testing of their Oral Polio vaccine. They have also successfully created a transgenic mouse which expresses the CD4 receptor found on the surface of white blood cells, the target of HIV virus. They are able to test drugs which interact with the CD4 receptor in mice instead of higher animals such as the chimpanzee, which possesses a CD4 receptor 98% identical to that of humans.

Industrial Applications

Transgenic animals can not only be created to produce products of pharmaceutical value, but also products of industrial value. An example is "BioSteel", the trademark name for the high strength fiber material made from recombinant spider silk-like protein which is produced in the milk of transgenic goats, made by a company called Nexia Biotechnologies. The protein is harvested from the milk, purified and transformed into microfibers. It is planned to use these fibres for bullet proof vests. In a paper published in Science, they describe the production of a number of different dragline spider silk proteins via cell culture techniques using silk genes derived from two different species of orb-weaving spiders. The silk proteins from one of these species were spun from a solution to produce BioSteel fibers, which were tested for a number of mechanical properties and compared to natural spider silk. They found they were able to produce fibers with mechanical properties similar to the natural spider silk including toughness, but lower strength. Toughness measures the amount of energy that the fibers can absorb before breaking and spider silks are among the toughest materials in the world.

Conservation and Ecology

In addition to the potential benefits of cloning of endangered species, biotechnology also plays a role in discovering, identifying and analysing genetic diversity in endangered species through the techniques of genetic characterisation of individuals in a population. Consequently, this could lead to regulation and conservation of genetic diversity.

The University of Pretoria, Department of Animal and Wildlife Science, is using molecular scatology as a technique for obtaining information on wild mammal populations. Molecular DNA analysis of faeces gives information as to the genetic structure and demographics of a population. Individual identification can be obtained through "DNA fingerprinting". This is particularly vital in animals that are endangered such as the cheetah, as it improves population management and conservation strategies.

The National Zoological Gardens has a genetic management research group which is focusing on genetic diversity in endangered or threatened species, including the extent and effects of inbreeding, the development of molecular markers for species identification.

The Agricultural Research Council Irene (Animal Genetics Division) has developed a National Genetic Database for game and Biobank for deposition and storage of biomaterials from a wide range of animal species. This will have various benefits in understanding population structure, the level of gene flow and genetic variation within and among populations and the level of inbreeding which will provide essential information for conservation management programmes. This will include rhino, sable, cheetah, leopard, lion, buffalo, kudu and more.

At the Indigenous Genotype Physiology & Biotechnology Development of the Agricultural Research Council (ARC)-Livestock Business Division (LBD): Animal Production Institute (API) research is being conducted on conservation of South African indigenous endangered breeds, through cryopreservation of semen, ova, embryo's and somatic cells. Animal biotechnologies such as transgenic and nuclear transfer are part of the longterm objectives of the group.

Another objective of animal biotechnology is to reduce the environmental impact of livestock farming. Transgenic animals are being created to improve the use of dietary phosphorous to lessen the environmental impact of animal manure. The Enviropig is the trademark of a genetically engineered pig developed at the University of Geulph that is capable of digesting plant phosphorous more efficiently than normal unmodified pigs. They produce the enzyme called phytase in their saliva which is active in the acid environment of the stomach and this breaks down the indigestible phytic acid into phosphate which is easily digested by the pig. Grains such as corn, soybean and barley contain 50 to 70% of phosphorous in the form of phytic acid which is indigestible to normal pigs, thus they require phosphate supplements in their diet which increases the phosphorous content of the manure. In heavy rain the phosphorous in the manure may leach into water in ponds, rivers and streams which facilitates algal growth, resulting in reduced oxygen concentration in the water and a death of fish and other aquatic animals. Enviropigs do not require phosphorous supplementation and excrete less phosphorous, reducing the potential of water pollution.

Another example of the impact of biotechnology on ecology is the production of infertile fish. Some species of fish being farmed are not indigenous to the area and can pose an ecological threat if they were to enter the natural ecosystem. Techniques have been developed to alter the chromosome composition of the fish to make them infertile and thus reduce any risk.

Animal Cell Biotcchnology

With the increasing world population, human being is posed with the greatest challenge of poverty reduction and feeding the population by increasing food production. In developing and poor countries, most of the population is dependent on agriculture and livestock production for their livelihood. Animal cell biotechnology can help this section of population to increase their production and come out of poverty. Artificial insemination has single handily played an important role in genetic up-liftment of inferior animal breed. The semen of high quality progeny tested bull can be disseminated all across the globe and thus directly change the genetic constitution of livestock. In addition, techniques like estrous synchronization, multiple ovulation embryos transfer (MOEF), In Vitro fertilization, sperm and embryo sexing etc. can make dairy and meat industry a highly profitable venture. Cryopreservation of semen and embryo has made it possible to store precious semen or high quality embryo for long time. The transportation of high quality animals from one part of world to other can be done very easily by this method. This technique is also significant in preservation of endangered species and thus preventing their extinction. Recombinant DNA technology has made it possible to produce the recombinant protein, hormone, enzyme or biopharmaceutical in bulk amount at lost cost. This has revolutionized the field of disease diagnostics, treatment and vaccination. Animal cell culture can be used for growing viruses in bulk amount which can be used for vaccine production. These finite and continuous cell lines can also be used for testing of toxins, chemicals, cosmetics and other unknown molecules. At this point of time, we can say that animal cell biotechnology has wide scope and efforts need to be made to make the technology cheap and affordable.

Future Perspective of Animal Cell Biotechnology

Animal cell biotechnology seems to have a bright future prospective. With the decrease in land holding and increase in population, there is increased pressure of enhanced livestock production from the limited resources. With the use of techniques like animal cell culture, artificial insemination, recombinant DNA technology, multiple ovulation embryo transfer, cryopreservation; it will be possible to increase the milk, meat, wool and leather production. Animal cell culture is used in molecular biology and cell for research studies. The future prospective of animal cell biotechnology is as follows:

Model System

The animal cell culture can be used as a model system to study about basic science like study of cell biology or biochemistry. It can be used to study about interaction between the types of cells and the agent which causes disease. The cell culture can also be used to study the drug effects and can also be used to study process of aging. Animal cell culture holds the potential of replacing excess of animal experimentation. It will relieve laboratory animals from pain and suffering, thus achieving animal welfare truly.

Cancer Research

Animal cell culture can be used to study about the difference between the cancerous cells and normal cells. Both types of cells normal and cancerous cells can be cultured in laboratories. With the help of chemicals, viruses and using radiation, normal cells can be transformed or normal cell can be converted into cancer cells. In this way, cell culture can be used to screen for antineoplastic drugs which can selectively destroy only the cancerous cells.

Increased Production of High Quality Animals

Superior genetic quality animals can be produced in large number by using multiple ovulation embryo transfer technology. High quality female is induced for super-ovulation using combination of hormones. These ova can be collected with the help of ultrasound and allowed to fertilize in laboratory by IVF technique. These embryos can be transferred to inferior quality surrogate mother that has been estrus synchronized. By this method, a high quality female animal can give birth to thousands of off-springs which is not possible naturally.

Desired Sex Offspring

A dairy manager always wishes for birth of female calf as it will increase the production capacity of the farm. With the help of sperm sexing technique, it has become possible to have offspring of desired sex. This technology hold potential for dramatically increasing the milk yield of dairy farms. It will also solve the problem of excess number of male offspring.

Virology

In virology, animal cell culture can be used for multiplication of the viruses instead of using animals. It can also be used in isolation and detection of viruses. Now, it may be possible to study the growth of viruses and the mechanism of virus replication.

Toxicity Testing

Animal cell culture can be used for checking the toxicity of new drugs and to study of cosmetics and different types of chemicals. Kidney cells and liver cells are most of the times used for toxicity testing. Animal cells cultured can also be used in determination of the new drugs dosage.

Vaccine Production

It is important to produce new vaccine for different types of diseases and animal cell culture can now be used for virus production. These viruses are useful in vaccine production. The types of vaccines such as vaccine for different types of diseases like for rabies, polio, chicken pox, hepatitis B virus (HBV) and measles.

Artificial Insemination

Artificial insemination is one of most powerful animal cell biotechnology tool that has increased the milk and meat production significantly. Artificial insemination has enabled the animal handlers to harness the capability of a superior bull to its maximum potential. Semen of a single superior bull can be used for insemination of thousands of buffalo. Artificial insemination in combination with MOEF can upgrade the genetic constitution of inferior animals. These techniques allowed mankind to envisage about a world of superior, high producing animals only.

Genetically Engineered Protein

With the help of recombinant DNA technology, the target gene can be expressed in specific host. This recombinant protein can be purified. A number of animal proteins/hormones/enzymes (Insulin, somatotrophin, interferon, trypsin etc.) have been produced using genetic engineering.

Replacement of Tissue and Organs

Animal Cell culture can now be used for replacement of organ or tissue. These replacement methods can be used for production of artificial skin for treatment of patients those having ulcers and burns. Various methods have been standardized for skin but there is urgent requirement to develop protocols for artificial organs like liver, spleen, kidney, heart etc. The number of donors for these organs is very less as compared to the recipients. Further refinement in animal cell biotechnology methods may make it possible to synthesize these organs in laboratory. The concept of humanized organs in related animals (i.e. pig, monkey) is also getting attention of scientific community. This method will be used for production of human organs in animals.

Gene Therapy

Animal cell biotechnology has also opened the gates of gene therapy. The cells are removed from the patient, missing a functional gene or lacking of a functional gene. These non-functional genes can be exchanged by the functional gene. These manipulated cells can now be inserted into the patient. The other method is by using the virus vector. In the genome of virus vector, a functional gene can be introduced and then this altered viral vector is inserted in the patient lacking functional gene.

Advancement in Human Heath

For human heath, animal made pharmaceutical can be used. Genes for important nutrients like Vitamin D3 can be expressed in dairy animals and the milk will be fortified with Vitamin D3. In the similar way, other nutrients/ biopharmaceutical genes can be specifically expressed in target tissue. This will supplement a healthy human society.

Animal Cloning

Animal cloning is currently an inefficient technique and only 1-3% of the nucleated egg cells develop into the live offspring. Various times the pregnancy loss has been observed just after the insemination of eggs containing adult cell nuclei into recipient. It has been reported that abnormalities in cardiovascular defect and in neurological system are more frequent in clones and many of these abnormalities appears from the incorrect epigenetic reprogramming. So it is necessary to focus in this area for better cloning. In the Dairy cattle we can focus increasing quantity for milk production in buffalo and cow. We should also focus to improve of body structure, enhancement of life-long productivity. In the beef cattle we can think about improvement of meat quantity and quality. For swine, we should focus in the improvement of daily weight gain and intramuscular fat. In Poultry, egg shell strength improvement and to decrease the fat volume inside abdomen should be focus.

Challenges

Although animal cell biotechnology has provided correct impetus to the scientists for welfare of animals and human being but these latest technologies need to be made affordable and feasible. Therefore, following area need to be strengthened in order to harness maximum animal cell biotechnology potential:

- Animal cloning and transgenic animal technique.
- Impact on human health and environment.
- How to yield very high quantity of milk, meat, and wool.
- Disease resistant animals.
- Use of animal for growing human organs.
- Production of hormones and growth factors in animal cell culture.

Biosafety Issues in Animal Cell Biotechnology

Bio-safety level is a discipline addressing safely handling and containment of the infectious microorganisms and hazardous biological materials. A bio-safety level is a set of bio-containment protection required to isolate dangerous biological agents in laboratory facility. There are four bio-safety levels of containment from the lowest to highest (BSL-1-4). Containment means managing the infectious materials in laboratory environment where they are being handled or maintained. Purpose of containment is to reduce or eliminate exposure of laboratory workers, other persons & environment to potentially hazardous agents. Appropriate facility design manner through engineering features, safety equipment and management practice should be laboratory for personal and environment safety. Animal and human cell cultures are very beneficial for diverse applications in modern biotechnology and biomedical research. Initially animal and human cell cultures used as substrates to produce viral vaccines. There are many applications of animal and human cell cultures to produce many products, like biopharmaceuticals, growth hormone, immune-modulatory, monoclonal antibodies and other medically important substances. Animal and human cell cultures can play very important role to studying various biochemical pathways, virus isolation from clinical samples, pathological mechanisms or intra- and intercellular responses. Due to enormous use of animal cells, there are large numbers of animal cell culture laboratories working in our country. There may be some risk to the human health and environmental due to these laboratories. All the work should be done under designated bio-safety level to minimise the health and environmental hazards. The risk assessment should result in the implementation of biosafety level measures while experience and practices provide maximal protection of human health and environment Along with the increasing importance of the contained use of animal and human cell cultures, biosafety concerns have pointed to the risks with respect to human health and environmental considerations. Risk assessment of animal cell cultures is based on both the intrinsic properties of the cell culture - including subsequent properties acquired because of genetic modification and the possibility that the cell culture may inadvertently or deliberately be contaminated with pathogens.

Why Biosafety Issues in Animal Cell Biotechnology?

- Laboratory Safety: Laboratory staff may be infected with some infectious pathogen when they will not follow biosafety rules.
- Blood-borne pathogens (BBP): Bio-safety level act as the barrier between laboratory staff and infectious pathogen which transmitted through blood or other body fluids.
- Recombinant DNA (rDNA): Bio-safety level prevent the spreading of recombinant DNA from the experimental cell to other living cells.
- Biological waste disposal: Bio-safety level provide safest way for the biological waste material disposal.
- Infectious & diagnostic specimen: Bio-safety level help in the safely handling of infectious and diagnostic specimen during the diagnostic procedure.
- Respiratory Protection: Bio-safety level act as the barrier between laboratory staff and infectious pathogen which transmitted through respiratory route.
- Bioterrorism: Bio-safety level act as the barrier between persons and material which have capacity to be utilization the bioterrorism. We can safely handle such material into the designated bio-safety level.
- Occupational safety and health in the use of research animals: Appropriate bio-safety level is helpful for the occupational safety and health in the use of research animals.

- Biohazards used in animal models: Appropriate bio-safety level requires for biohazards used in animal model system during various experiments in biological sciences.

Bio-Hazardous Materials: May be Present in Cell-Line

While working with animal cell culture, there may be presence of adventitious contaminating agents constitutes the main hazard to humans. Contamination may occur by the source (e.g. infected animals or tissues), during cell handling (repeated passages and use in the laboratory) or by using contaminated biological reagents (e.g. media and additives derived from bovine sources are often contaminated with bovine viral diarrhoea virus, BVDV).

- Bacteria and Fungi: Sometime bacterial or fungal contamination can be readily detected in cell cultures because of their capacity to overgrow cell cultures. Bacterial and fungal infections are relatively easy to prevent and to treat. The source of bacterial or fungal contamination may environment and persons working in culture room.
- Mycoplasma: Compared to bacterial or fungal infections, contaminations with mycoplasma give more problems in terms of incidence, detectability, prevention and eradication. Mycoplasma infection may go undetected for many passages, causing a variety of unpredictable effects causing harm to the host cell. Mycoplasma infection may also influence the sensitivity of host cells for growth of viruses.
- Viruses: Viral contamination needs attention because infection may be without cytopathic effect for the cell culture or may be latent (e.g. herpesvirus) and hard to detect. Human and non-human primate cultures are more likely to harbour viruses that are highly pathogenic to humans. Of concern are the blood-borne viruses such as Hepatitis B Virus (HBV), Human Immunodeficiency Virus (HIV) and others such as Hepatitis C virus (HCV) and Human T-cell lymphotropic viruses (HTLV). However, non-human cell cultures are not without risks as they may contain viruses with a broader host range able to infect humans such as rodent cell culture carrying Hantavirus or primate cells harbouring Marburg virus.
- Parasites: Adventitious contamination with parasites may be an issue when handling freshly prepared primary cell cultures or tissue cultures originating from a donor organism that is known or suspected to be infected with a specific parasite. Well-known intracellular protozoan parasites for which laboratory-acquired infections have been reported are Toxoplasma gondii, Trypanosoma cruzi, Leishmania sp, Cryptosporidium parvum, Plasmodium sp. etc.
- Prions: Though a limited number of cultured cell lines (e.g. mouse neuroblastoma cell lines Sc N2a) have been shown to promote, upon sub-passaging, stable and persistent replication of PrP(Sc) as well as infectivity, most cell lines are resistant to prion infection. However, in contrast to most of the infectious agents, prions are particularly difficult to inactivate. In fact, no method can guarantee total inactivation of these agents. So, one should bear these considerations in mind when using growth media of bovine origin.
- Other Cell-line: A cell-line may be contaminated with other types of cell-line while animal cell culture laboratory maintaining more than one cell-line simultaneously. Contamination with other types of cell-line will affects the results of cell-line.

Principles of Biosafety

Biosafety guideline provided the safely handling of bio-hazardous at justified cost. Biosafety guideline helpful for increasing levels of worker and environmental protection. Biosafety guidelines are compulsory part for working safely in research & medical laboratory facilities.

General Good Lab Technique

- Hygienic Practices,
- No Smoking in lab,
- No Eating in lab,
- Do-not store the food in lab,
- Do-not applying cosmetics, lip balm, contacts lens,
- Restricted entry,
- Wash hands after procedures,
- Proper discard of waste,
- Decontaminate lab bench before and after work,
- Proper clothing,
- Plan your work,
- Read Material Safety Data Sheet (MSDS).

Standard Microbiological Practices

- Most important follow SOP,
- Awareness of potential hazard,
- Trained & proficient in techniques,
- Supervisors responsible for:
    - Appropriate laboratory facilities,
    - Personnel & Training,
    - Special practices & precautions,
    - Occupational Health Programs.

Different Biosafety Level

Biosafety Level-1 (BSL-1)

- Well characterized agents.

- Agents not known to cause disease (in the healthy human adults).
- Prophylactic treatment available.
- Open bench procedures.
- Animals in open cage system or open environment (out-doors).
- Good laboratory practices.
- Example: E. coli, Transgenic Plants, Plasmids, Fungi, Mold and Yeast.

Basic Instruments

- Bench-top work allowed.
- Daily Decontamination.
- Manual pipetting.
- Required Hand washing.
- Red bag waste.
- Bio cabinet not required (unless creating aerosols).

Biosafety Level-2 (BSL-2)

Practices and Procedures

- Agents associated with human disease.
- Treatment for disease available.
- Agent poses moderate hazard to personnel & environment.
- Direct contact or exposures are dangerous.
- Percutaneous exposure like scratch, puncture, needle stick.
- Mucus membrane exposure for example eyes, mouth, open cut.
- Example: HSV, Hepatitis viruses, Dengue.

Basic Instruments

- Limited access to laboratory when work in progress.
- Daily decontamination.
- Mechanical pipetting.
- Lab-coat, safety glasses and gloves required.
- Red bag & sharps containers required.

- Bio-hazdous Sign posted at the entrance to laboratory.
- Label all equipment properly (incubators, freezers, etc.).
- Tissue Culture room with negative air flow.
- Documented training of staff.
- Baseline serology or pre-vaccination may be required.

Biosafety Level-3(BSL-3)

- Indigenous or exotic agents.
- Aerosol transmission.
- Serious health effects.
- Treatment may or may not exist.
- Example: Human Immunodeficiency Virus (HIV), Mycobacterium tuberculosis (TB), Coxiella burnetii.

BSL-3 Practices

- Public access NOT permitted.
- Daily decontamination after spill and upon completion of experiment.
- Autoclave required and waste is disposed at the end of day.
- Required foot activated hand-washing sink controls.
- No sharps unless absolutely necessary.
- Aerosol minimization procedures required.
- Wrap around disposable clothing is required.
- Specialized equipment may be required depending upon procedures.
- Bio-hazdous signs and labels posted and Air flow control.
- Documented training and personnel competency certification.
- Base-line serology.
- Spills-report immediately and treat accordingly.
- Vaccinations/post exposure protocols.
- SOP, Biosafety Manual, Biosafety Officer should present in lab.

Biosafety Level-4 (BSL-4)

- Builds on BSL-3/ABSL-3 practices.

- Maximum containment facilities.
- Pressurized Containment Suite.
- BSL-3 + Class III Biosafety Cabinet.
- Chemical decontamination showers.
- Liquid effluent collection decontamination Facility.
- No BSL-4 labs exist at MDU.
- Example: Lassa Fever Virus, Ebola Hemorrhagic Fever Virus, Marburg Virus.

Transcriptomics and Proteomics

The term transcriptome embraces all types of complementary RNA synthesized during transcription of the genome at a given time point. Of these, microRNA (miRNA) and messenger RNA (mRNA) momentarily are the RNA types of greatest interest in transcriptome research. The mRNA expression levels directly reflect the gene activity of a cell, while differences in miRNA expression levels, a major posttranscriptional regulator of gene activity, are thought to be associated with several cellular functions and diseases. Three common methods for the multiplex detection and quantification of miRNA and mRNA at a large scale are momentarily used: microarrays, deep sequencing (RNA-Seq), and quantitative reverse transcription (RT) PCR arrays.

Table: Advantages and limitations of transcriptomics technologies.

	qPCR arrays	Microarrays	Ref-Seq
Advantage	Highest sensitivity/specificity.	Costume-made assays available less expensive than Ref-Seq.	Combination of quantitative and sequence information.
Limitation	Only few hundreds of targets measurable in parallel.	Less sensitive and specific than qPCR.	Less sensitive and specific than qPCR Expansive.

Microarrays

A microarray is a multiplex, high-throughput screening method which uses a two-dimensional arrangement of nucleotide probes on a glass slide or a thin silicon chip. One array can contain ten thousands of nucleotide sequences or hypothetically the complete transcriptome, which is covalently bound to the chip's surface or fixed in parallel rows in microfluidic channels. Depending on the scientific questions asked, these nucleotide sequences may represent short fragments (probes) with exon sequences to analyse the presence and the quantity of specific mRNA or miRNA in expression profiling experiments. Alternatively, genomic DNA sequences with allelic variants, known mutations, or single nucleotide polymorphism (SNP) is spotted on the chips to identify the mutational status of a tissue in comparative genomic analysis or SNP assays. Independent from the spotted sequences, pairwise hybridization of the given sequences on the chip with complementary DNA or cDNA sequences (targets) in a tissue lysate under high-stringency conditions is the general principle of all microarray-based experiments.

Successful hybridization is visualized by labelling of bound target oligonucleotides with fluorescent dyes. Fluorescence intensity of the respective spot on the chips is subsequently used to determine the presence or absence of a mutation or SNP or the level of mRNA or miRNA expression in a tissue, respectively.

The resulting large data sets on the expression level of up to ten thousands of genes represent a challenge to the commonly used statistic methods. These challenges include the subtraction of background noise, the normalization of data to allow for the comparison of different microarrays and microarray experiments with each other, and the identification of statistically significant changes in large data sets. For the latter, the common statistical test, *t*-test, ANOVA, and Mann-Whitney test have been adjusted to consider the multiple comparisons and permit cluster analysis. However, in most instances, the number of variables is larger than the observational units. Principal component analysis and partial least squares have been used for dimension reduction and to visualize microarray data sets. Furthermore, clustering methods like hierarchical and fuzzy clustering are used to identify differences in gene expression profiles associated with different disease states or treatment response.

Ready-to-use whole genome cDNA microarrays are commercially available for several species including human, mouse, and rat (Affymetrix, Illumina, Agilent) but also for canine, bovine, porcine, sheep, equine, chicken, rhesus macaque, salmon, and zebrafish (Affymetrix, Agilent). However, costume-made microarrays with any given DNA of interest are provided by several local and international suppliers. These arrays can be adjusted to each experiment in terms of gene number or type of interest and may therefore represent a low-cost alternative to expensive full-genome microarrays.

Similar to other analytic tools, microarray experiments require technical and biological replicates. In addition, several initiatives are launched to standardize microarray experiments, whose outcome is strongly influenced by tissue and RNA handling, the assay protocols and the microarray platform on the resulting data. For instance, the "minimum information about a microarray experiment (MIAME)" and the "microarray quality control (MAQC) project" are becoming the gold standard of how detailed information on an experiment should be included in a publication.

Microarrays are well established and valuable tools to get a first impression on the genetic activity or the genomic composition of a specific tissue under physiologic and pathologic conditions. Despite the high quality of prefabricated microarrays, one has to keep in mind that cross-hybridization and low sensitivity when compared to quantitative PCR are intrinsic problems of this valuable method and may lead to false positive or negative results. Microarray data can therefore in most circumstances not be considered as final and conclusive information but have to be confirmed by other methods such as quantitative PCR or, optimally, on the protein level or by functional cell assays.

Deep Sequencing (RNA-Seq)

RNA-Seq is a new approach to obtain complete information on the RNA expression levels in a tissue sample. This approach utilizes the tremendous progress in the development of next generation sequencing technologies, which allow to sequence millions of base pairs in a relatively short time.

Initially, these methods were used for whole-genome sequencing of organisms or tumours. Within the past 2-3 years, second-generation sequencing has been applied to cDNA sequencing to obtain information on the full set of all transcribed RNA sequences within a given tissue, that is, the complete transcriptome in so-called RNA-Seq experiments.

RNA-Seq uses a cDNA library, which is reverse transcribed from the total RNA in a single tissue or cell culture. Small fragments from each of these cDNA molecules are then sequenced using the different technology platforms Illumina IG, Applied Biosystems SOLiD, Roche 454 Life Science, or Helicos Biosciences tSMS.

Although subsumed under the name "next generation" or "deep" sequencing, the three main technologies are based on different principles. Pyrosequencing, which has been developed by 454 Life Sciences, is based on the emulsion amplification of a single DNA sequence attached to a primer-coated bead, on which a clonal DNA colony is formed in a picotitre well. The sequencing procedure employs luciferase to generate light for indication of the addition of one of the four consecutively added dNTP types to the newly synthesised DNA in each well.

In contrast, the reversible dye-terminator-based technology of Ilumina employs amplification of DNA fragments to clonal DNA colonies on a glass slide. During each sequencing step, one of the four types of dNTP is added, and the addition of the fluorescently labelled nucleotides is detected by a camera at the respective spot. After each step, the dye is chemically removed, and the next cycle is started.

The SOLiD technology of Applied Biosystems is based on sequencing by ligation. Similar to pyrosequencing, DNA is amplified by emulsion PCR, and the resulting bead with a clonal DNA colony is positioned on a glass slide. The mismatch sensitivity of the DNA ligase is then used to identify the nucleotide present at a given position in a DNA sequence.

The sequences retrieved from any of these technologies are aligned to the reference genome, and the resulting information can be used to analyze both genomic and posttranscriptional mutations and the mRNA expression level of each gene. RNA-Seq therefore extends the information obtained with microarrays by a second dimension, the sequence of the mRNA transcripts. This may permit the identification of disease-associated mutations or SNP in the actively transcribed genes.

Another advantage of RNA-Seq is potential analysis of genomes from organisms without complete genomic sequence information, although further data analysis in these cases is often hampered by the lack of gene annotation or the lack of a sequenced genome. This is in strict contrast to the microarray approach that requires knowledge on the sequences to be detected. Finally, RNA-Seq also has a large dynamic range of quantification levels which allows the quantification of very low and high quantities in contrast to the low sensitivity and reduced dynamic range of microarrays.

A major disadvantage of this technology is the relatively high costs for a high coverage analysis of a complex mammalian genome. Still, the price per nucleotide is significantly lower than for classic sequence; a reasonable RNA-Seq run requires the analysis of several million bases, making it still an expansive technology for most scientific questions.

Quantitative Real-time PCR Arrays

Quantitative real-time PCR (qPCR) arrays are, although less complex, still a very valuable method for the quantification of the gene expression of up to 600 genes. At the moment, qPCR arrays are a commonly used method for the quantification of the "miRNA transcriptome". This is mainly based on two facts. First, the number of miRNA types per species can be considered at level of hundreds and is therefore much lower than the number of mRNA types. Second, the high similarity between the different miRNA types makes the development of microarrays difficult due to the high risk of cross-hybridization. In contrast, Taq-Man- or SYBR-Green-based qPCR assays permit a specific and sensitive quantification of miRNA and are a commonly used method in veterinary sciences for the detection of mRNA and miRNA. Arranged in an array, qPCR assays are therefore an efficient and highly reproducible method to quantify the miRNA transcriptome.

Methods in Proteomics

Transcriptome analysis gives useful information on the transcriptional activity of a cell. However, due to posttranscriptional gene regulation and the different stabilities and biological half-lives of RNA and protein specimens including miRNA, the correlation between mRNA levels and the corresponding protein is often poor. Transcriptome analysis therefore requires confirmation of the results on the protein level for most scientific questions. The analysis of the cellular proteome is, however, a much greater challenge than the transcriptome due to the high chemical diversity of the cellular proteins. Despite these difficulties, proteome analysis has become an important tool for explorative analysis of molecular mechanisms in physiology and pathology. There is a great variety of approaches applied in proteome research that can be grouped into two broader categories: gel-based assays and shotgun proteomics.

Table: Advantages and limitations of the two major proteomics approaches.

	2D-DIGE	Shotgun Proteomics (MudPit)
Advantage	Moderate costs for necessary equipment (separation phase) Good protein separation.	Industrialized work processes Almost all protein types analyzable.
Limitation	Gel-based variability Hydrophobic (membrane) proteins difficult to measure.	Difficult quantification of low abundance proteins in complex protein mixes High costs for equipment.

Gel-based Proteomics

One of the major problems of proteome analysis is the complexity of proteins in tissues and cell lysates. Two-dimensional gel electrophoresis (2D-GE) which combines two dimensions of physical protein separation is a common method to separate proteins by their chemical properties. In the first dimension, the proteins are separated by their isoelectric point (IP), that is, they are arranged on a linear gel with an immobilized pH-gradient according to their content of basic or acidic amino acids. In the second dimension, IP-separated proteins are separated by their molecular size similar to a conventional SDS-Page. The orthogonal combination of both techniques results in a two-dimensional arrangement of proteins with a high resolution of single protein types.

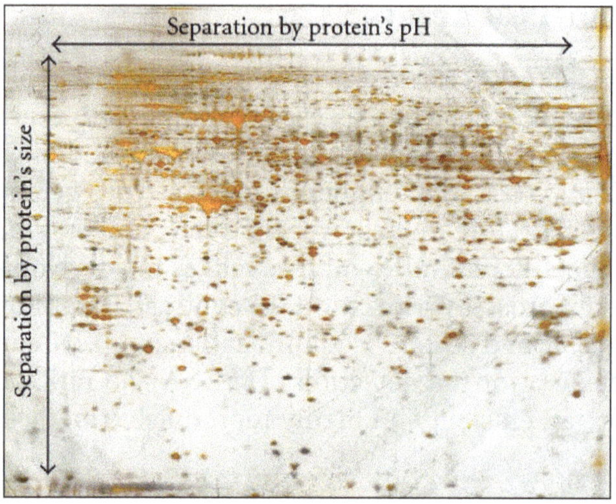

Figure shows gel-based proteomics reduces the complexity, of protein lysates by two-dimensional separation of proteins, that is, separation by size and pH. This leads to a high resolution of single protein which facilitates the quantification of their spot size and the extraction of single proteins for subsequent mass spectrometry-based identification of the protein identity.

The separated proteins can be visualized on the gel by different staining methods, for instance, Compassion brilliant blue, silver, or fluorescent stainings. Each of the detected spots on the acrylamide gel theoretically consists of one protein species, and each cell or condition of a cell is theoretically associated with a specific pattern of these spots on the gel. Changes in this spot pattern should therefore reflect changes of the cellular proteome, for example, its metabolism and gene activity under diseased or healthy condition.

Two approaches have been developed to compare spot presence and intensity in 2D-GE gels. In the original single channel approach, all gels containing one protein sample are stained with same stain and scanned independently. This approach is, however, prone to the technical intergel variability and has difficulties to normalize spot intensity. The multiplex approach of two-dimensional differential gel electrophoresis (2D-DIGE) uses gels which contain up to three protein samples stained with three different fluorescent dyes to reduce the gel number. One of the three samples on each gel usually represents an internal standard, which allows for normalization of spot intensity between different gels.

Detection of the differences in spot intensity even between only two gels usually overstrains the human eyes capacity independent from the method used. Comparison or quantification of spot intensity and presence therefore requires automated digital image analysis. Despite the high efficiency of the available software, comparison of protein expression pattern on polyacrylamide gels is still the major challenge and weakness of gel-based proteome analysis. Even with highest accuracy and long experience of the experimenter, preparation of 2D-GE gels without technical artefacts is not assured mostly due to random polymerization artefacts in non-manufactured gels. Commercially available, machine-made gels are an interesting but costly alternative that reduces most of these gel-based artefacts.

Once protein spots with a significant difference in intensity between the two disease states are identified, the protein type has to be identified using mass spectrometry. To this end, the spots with the respective protein are picked and eluted from the gel and fragmented for mass spectrometry by trypsin-digestion. The molecular weight of the trypsin digested protein fragments is then determined using, for instance, Matrix-assisted Laser Desorption/Ionization (MALDI). This approach uses trypsin fragment pattern, the so-called "peptide fingerprint" (PF), to identify the regulated protein in open source protein databases. With this PF, protein data bases are searched to determine the peptide/protein with the highest accordance according to statistical calculations. However, PFs for animal proteomes are far from being complete, and PFs obtained from proteome studies in animals are in most cases compared with peptide masses of human protein in the versatile human data bases. This may lead to low sequence coverage, and peptide sequencing is required for Mascot MS/MS Ions Search.

Gel-based proteomics is still an acceptable and commonly used approach to analyze the cellular proteome. Its main advantage is the relatively low costs for equipment when compared to non-gel-based approaches. Disadvantages are the restriction on hydrophilic proteins with a pH between 3 and 10, difficult standardization of the gel preparation, the limited dynamic range, and the limited sensitivity.

Shotgun Proteomics (MudPIT)

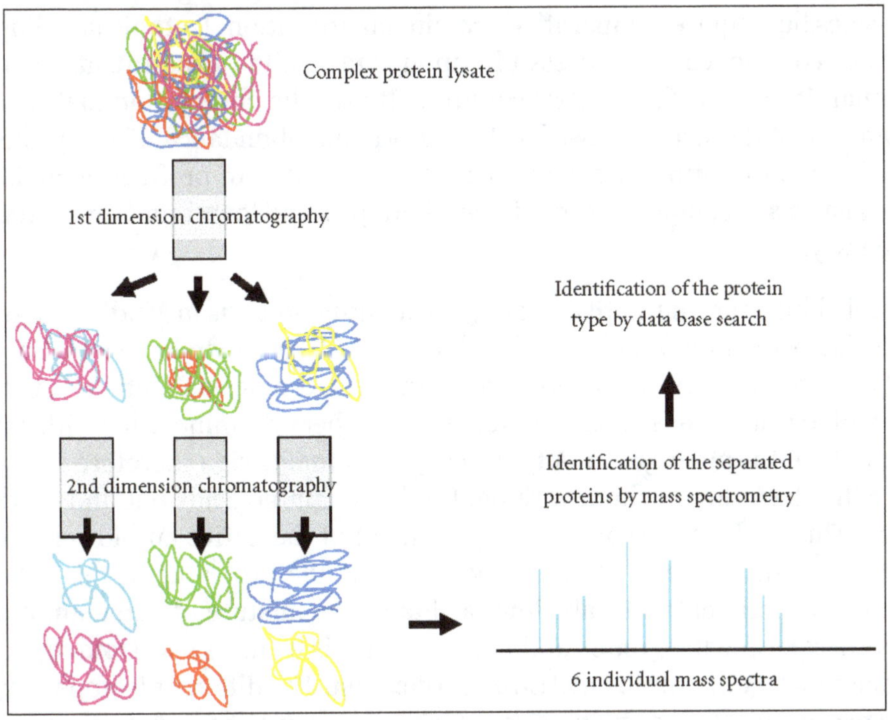

Figure shows shotgun proteomics or MudPIT uses multidimensional chromatographic separation to reduce complexity of protein lysates. The selection of the type of chromatographic columns depends on the question asked and the technical equipment. The different protein lysate fractions obtained by chromatographic separation ideally contain only one protein type which is subsequently identified by mass spectrometric analysis and data base search.

The term shotgun proteomics is applied due to its similarities with DNA shotgun sequencing, which sequences multiple short DNA fragments in complex mixtures and recombines them in silico. Similarly, shotgun proteomics analyses trypsin-digested protein fragments in complex mixtures by mass spectrometry after separation by liquid chromatography. This so-called multi-dimensional protein identification technology (MudPIT) combines at least two chromatography separation steps with tandem mass spectrometry. The combination of two chromatographic methods allows for the separation of digested peptide fragments by at least two features, for example, charge and hydrophobicity, to reduce the overall protein complexity. The selection of two coupled, thus two-dimensional chromatography methods depends on the scientific question asked and the experience of the experimentator. Microcapillaries packed with strong cation exchange material, reverse phase material, or other affinity-based chromatographic materials are most commonly used. In addition, separation before mass spectrometry may also include isoelectric focussing and capillary electrophoresis which enables a high resolution and narrow analyte bands.

The different peptide fractions are eluted from the chromatography columns by alternating salt pulses or organic mobile gradients. They are subsequently stored or, more commonly, directly applied to the mass spectrometer. MudPIT approaches commonly use electrospray ionization (ESI) mass spectrometers since the chromatography eluate can be directly "online" analyzed with a connected ESI mass spectrometer.

Most proteome experiments do not exclusively aim at identifying the general protein composition in the tissue lysates but require comparative protein quantification. Isotopic labelling or label-free methods have been developed and successfully applied in MudPIT experiments. A simple method for label-free quantification is the spectral counting. It uses the observation that the total number of spectra from a peptide correlates well with the original abundance of the protein in lysates. However, relative quantification by spectral counting is limited to prefractionated, less complex samples, while analysis of complex, unfractionated samples and low abundant proteins may result in reduced accuracy.

Stable isotope labelling is an alternative strategy to quantify proteins in MudPIT experiments. Stable isotope labelling with amino acids in cell culture (SILAC) is a metabolic labelling strategy that has been applied to both cell cultures and small mammals. It relies on metabolic incorporation of "light", native amino acids in one experimental group or "heavy" amino acids with substituted isotopic deuterium ^{13}C in the other group. A typical experimental design therefore includes a cell line which is grown in cell culture medium with light or heavy amino acids but under otherwise identical conditions. This leads to the complete replacement of the native amino acids in the "heavy" isotope group after a decent incubation period, marking the proteome of this cell unequivocal. After the intended experiment, this labelling facilitates the mixture of equal amounts of protein lysates from both experimental groups and parallel analysis by mass spectrometry, thus reducing technical artefacts during the spectrometric quantification. The different isotope mass in the amino acids of the two groups leads to twin peaks which can be quantified relatively according to their intensity ratios.

Isotope-coded affinity tags (ICAT) and isobaric tag for relative and absolute quantisation (iTRAQ) rely on the chemical labelling of protein samples after the experiment but before mass spectrometry. ICAT and iTRAQ assays label proteins with reactive groups which contain either heavy or light isotopes, which allows the direct quantification of proteins by mass spectrometry similar to SILAC. In contrast to ICAT and SILAC, the isobaric labelling by iTRAQ does not increase the complexity of the protein lysates and labelling of all peptides in a mixture after the biological experiment. ICAT and iTRAQ therefore enable the proteome quantification in tissue samples of larger mammals which cannot be fed with substituted amino acids only.

Shotgun proteomics analyses the proteome of complex protein lysates in an industrialized manner without gel-based artefacts. It permits the analysis of a wider range of proteins including hydrophobic membrane proteins and strongly acidic and basic proteins. Limitations are the restriction of this method to specialized and appropriately equipped laboratories and the relatively high costs.

The "-omics" technologies enable the parallel analysis of several thousands of biomolecules in a sample instead of the traditional sequential approach. The comprehensive parallel approach naturally interferes with the sensitivity and specificity of detection and quantification. However, these technologies are now on a technological level that facilitates their application for explorative, descriptive pilot studies on biological phenomena when the present knowledge is not sufficient to establish a substantiated hypothesis on its molecular basis. Furthermore, they may also be applied to overcome deadlocked views and paradigms and to approach well-known diseases unbiased again.

In the following, three major research fields are introduced where "-omics" technologies are successfully applied: biomarker discovery, identification of complex, prognostic gene expression patterns, explorative and hypothesis-driven proteome, and transcriptome analysis.

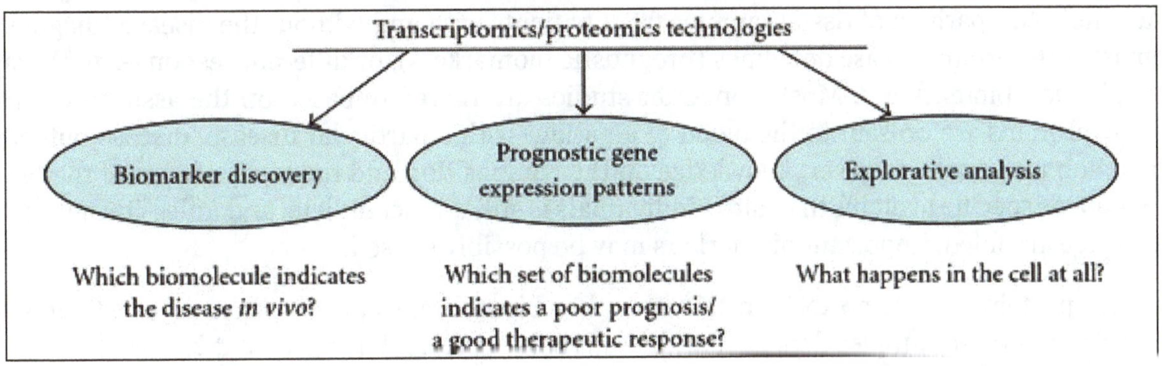

Figure: The three main fields for the application of transcriptomics and proteomics technologies and the basic questions behind these applications.

Biomarker Discovery

A biomarker can be DNA, mRNA, miRNA, proteins, or any other biomolecule associated with a specific state of the cell or tissue. These markers are synthesized by the diseased tissue itself or by nondiseased cells in response to the neoplastic or inflammatory disease. Biomarkers are further subdivided in diagnostic marker which enable the detection of the disease, prognostic markers which in turn allow for a prediction of the disease course when the initial diagnosis has been established, and stratification markers which predict a response to a specific treatment.

The ideal biomarker therefore permits the diagnosis and classification of a disease with a cheap, sensitive, and specific assay in a tissue sample which is obtained by a microinvasive intervention such as serum samples or small tissue biopsies.

Several, well-established biomarkers are used in clinical biochemistry in daily diagnostic routine. For instance, enzymatic activity of alkaline phosphatase (ALP) and other enzymes in serum samples are used for the evaluation of the hepatic function or thyroxin, and sex steroid levels are used to identify endocrine diseases. Tumor (bio-) markers are a field of specific interest in the search for biomarkers, and several of them have been introduced in human routine diagnostic, for instance, alpha-fetoprotein and carcinoembryonic antigen. In contrast, several, mostly immunohistochemical but also serum tumor markers have been suggested in veterinary medicine but are in most cases not routinely used in the prediction of disease outcome mostly due to low specificity and sensitivity.

The complete sequencing of the canine, feline, bovine, equine, and chicken genome, the availability of manufactured genome microarrays, and the reduction in costs of transcriptomics and proteomics technology will however accelerate the search biomarker in veterinary sciences, and progress in this filed can be expected in the near future. However, biomarker identification is, despite the technological progress, still difficult and requires structured and expansive analysis and comprehensive evaluation of the results.

The first and in most cases a tremendous hurdle for biomarker studies in veterinary medicine is the need for well-preserved, clinical samples with comprehensive clinical data on disease outcome. The appropriate number of samples from a well-defined study population taken with a standard protocol usually requires multi-centred efforts to gain the critical mass. Once a sufficient and statistically relevant number of clinical samples are available, biomarker discovery is usually based on a simple comparison of tissue samples from patients with and without the disease (diagnostic biomarkers), certain disease outcomes (prognostic biomarkers), or different responses to therapy (stratification biomarkers). Most biomarker studies are therefore based on the assumption that specific proteins are present in the blood of a patient with a particular disease, disease outcome, or reaction to a drug. However, knowledge on the composition and the biodynamics of the blood proteome or specific proteins in healthy individuals is still far incomplete, and unless these gaps of knowledge are filled, important biomarkers may be possibly missed.

Whether proteins, DNA, miRNA, mRNA, or other biomolecules are searched as candidates for biomarkers in these samples depends on the available samples the concurrent knowledge on disease mechanisms, and the general requirements of the intended final assay. Nonetheless, there is a clear trend towards proteomic approaches on easily accessible body fluids like serum to identify biomarkers for clinical routine diagnostics. This is supported by the growing content of the public protein data bases (SwissProt, UniProt, RCSB) for human, murine but also other farm and companion species which facilitates the first biomarker discovery studies in veterinary science.

The work plan for biomarker discovery projects is long and complicated. Briefly, the early detection research network (EDRN) has proposed a five-phase process for the development and testing of disease biomarkers.

Explorative preclinical, proteome or transcriptome studies are proposed as initial first phase steps in biomarker discovery. Which of the "-omics" strategies is utilized is in most circumstances determined by the available equipment and the amount of preexisting knowledge of molecular disease mechanisms. Nevertheless, mass-spectrometry-based profiling by MudPIT proteomics of low molecular weight proteins (peptidome) in the clinical serum samples has been given special attention due to the easy and microinvasive obtainment of such samples in a clinical environment. In contrast, the cancer biomarker-family approach is based on the hypotheses that "if a member of protein family is a biomarker, other members of that family might also be good biomarkers". Finally, the secreted-protein-based approach hypothesizes that a good serological biomarker should be a secreted protein, and biological fluids with contact to the tumor site may help in primary identification of these markers.

The initial preclinical, explorative phase of the biomarker discovery which includes proteome or transcriptome profiling of the complete set of proteins has to be followed by the development of clinical assays for the identified few candidates. The assays are tested retrospective longitudinally and prospectively to evaluate their sensitivity and specificity for disease detection in phases two to four. Finally, in phase five, the established assays have to be evaluated in a large patient population to ensure their clinical efficiency and impact, another phase which requires multicenter collaboration to obtain sufficient patient numbers. All of these stages are influenced by several variables that have been carefully planned or standardized by standard operating procedures but are beyond the scope of this review.

Costs are another critical factor of biomarker discovery at each of the discovery phases. In the initial exploratory phase, the high costs of modern instrumentations and reagents for proteome and transcriptome analysis are a limiting. Costs for commercial proteome analysis are dropping. In the proceeding phases, costs are mainly driven by the development and validation of marker-specific assays, for example, ELISA. The cost for the development of the assays varies but numbers of 100,000 up to 1,000,000$ have been suggested. In addition, experience shows that only a small fraction of the total candidates successfully make good clinical biomarkers. The high developmental costs for specific biomarker detection assays and only to a minor extent the costs for explorative "-omics" technologies are therefore considered as the current bottleneck in biomarker discovery.

Detection of Prognostic Gene Expression Patterns by Supervised Gene Expression Profiling

Especially in tumor biology, it becomes more and more evident that for most cancer types a malignant phenotype is associated with specific, complex changes in gene expression levels and not based on the expression levels of single genes. This perception has changed the search for prognostic factors in oncology. In recent years, expression profiling has been used to identify complex, prognostic gene expression patterns. Ideally, gene expression profiling focuses on protein expression levels, since mRNA or miRNA expression levels do not fully reflect the protein expression pattern and the biologic activity of a cell. However, the technological head start of mRNA assays like microarrays and next generation sequencer makes these methods superior to proteome analysis for the identification of prognostic gene sets. In addition, prognostic gene expression patterns do not have to reflect the true biologic state of the cell. Their only function is to diagnose a disease or predict a disease outcome even if they are based on biologically irrelevant, phenotypic mRNA patterns.

Two major approaches for gene identification are commonly used, supervised and unsupervised classification. The principle of supervised gene expression profiling studies is rather simple. According to the diagnostic questions, the complete mRNA of tissue samples from "training" groups of patients with and without the disease or the intended or not intended therapy outcome are analyzed to identify genes whose expression level is significantly associated with the respective group. Thus, the tissue samples are first divided by their disease status, and the data analysis aims at identifying the differences in the gene expression profile behind these clinically observed differences. In contrast, the unsupervised method differentiates tissue samples according to their gene expression profile without preceding information on their diseases status. This approach aims at identifying molecular phenotypes within a by then phenotypically homogeneous disease which may be of interest for treatment and prognosis. Another method is to apply experimentally obtained information on gene expression profiles associated with the activation of certain molecular pathways to classify tumour samples. This hypothesis-driven gene expression profiling approach aims at differentiating tumours that depend on these molecular pathways and may be treated accordingly.

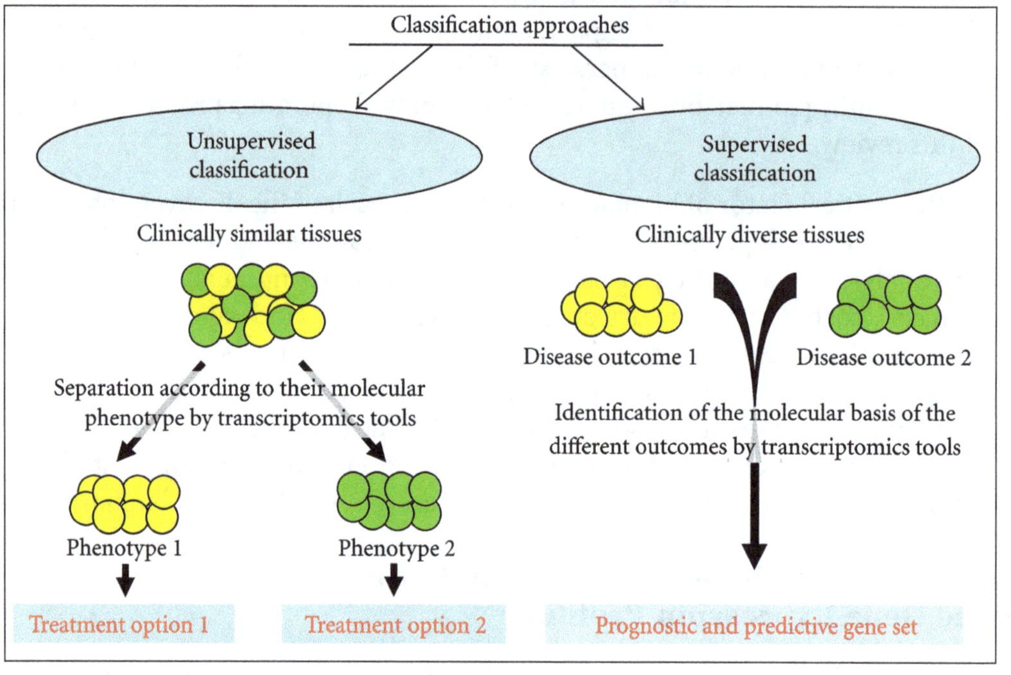

Figure shows two major approaches for the identification of disease-relevant gene expression profiles are supervised and unsupervised classification. Unsupervised classification is used to identify different molecular mechanisms in tissues from patients with clinically similar disease phenotypes. Clustering of these tissues according to their molecular phenotype may lead to the identification of different molecular phenotypes behind the identical clinical behaviour. This knowledge may finally lead to diverse therapy protocols of these molecular subtypes. In contrast, during supervised clustering, molecular mechanisms behind different disease outcome or phenotypes are analyzed. Two tissue groups with clinical differences are therefore compared, and the differences in gene expression are thought to be relevant for the distinct clinical behaviour. The resulting set of genes may also serve prognostic markers for the prospective classification tissue samples.

The diagnostic value of the gene sets identified with supervised clustering has to be tested in a second, independent "test" group to confirm its usefulness. This group again consists of a large population of patients with and without the disease or the different therapeutic outcomes. In an unsupervised approach, the prognostic gene set is now used to assign tissue samples to one of the groups. According to the ability to attribute clinical samples to the respective group, that is, false negative and false positive results, the specificity and sensitivity of the gene set are determined.

In 2002, van't Veer et al. published one of the first studies that identified a prognostic set of 70 genes for human breast cancer. In a supervised classification approach, they compared lymph-node-negative breast cancer samples with short intervals to distant metastases with samples with long intervals to distant metastasis. By this approach, the authors identified a prognostic set of 70 genes that predicts distant metastases and allows a patient-tailored therapy strategy to identify patients which benefit from the application of adjuvant therapy from. Of interest, first gene expression profiling studies on canine mammary tumours show that similar prognostic gene set may exist for malignant canine mammary tumours although their prognostic relevance has not been tested in larger patient populations yet. It can however be assumed that similar prognostic gene sets will be identified for veterinary diseases in the near future, given that affordable molecular tools like microarrays are now available for the most relevant species in veterinary science.

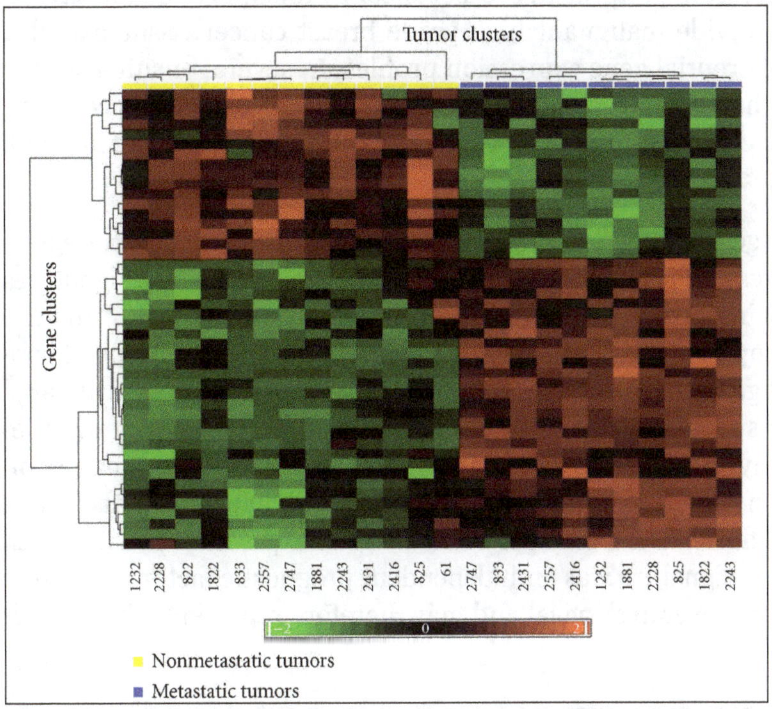

Figure: First studies in canine mammary tumours show that clinically different but histologically similar tumours can be clustered in a supervised approach into clinically relevant groups by a prognostically relevant gene sets.

Unsupervised cluster analysis and hypothesis-driven gene expression profiling are less successful and promising strategies to identify prognostic gene expression profiles but may help to understand disease-associated molecular mechanisms in explorative and hypothesis-driven transcriptome and proteome studies.

Explorative and Hypothesis-Driven Transcriptome and Proteome Analysis

Ideally, scientific progress is a continuous hypothesis-driven process by which new ideas are based on and refine recent findings. However, occasionally the preexisting knowledge is insufficient to establish a substantial hypothesis. In these cases, transcriptome and proteome analyses are a promising, explorative approach to obtain first exploitable information on potential molecular disease mechanisms. Both supervised and unsupervised gene expression clustering using microarray technology and comparative proteome analysis have been used in veterinary science to explore the molecular basis of various, mostly neoplastic diseases. All these studies compared diseased and non-diseased tissues in a supervised approach to identify genes or molecular pathways with significant disease association. These disease-associated proteins are now to be tested in further hypothesis-driven experiments to prove their relevance as primary disease causes or to identify them as pure phenotypic and secondary effects.

Unsupervised clustering of microarrays data has been used in human oncology to sub-classify histologically identical cancer samples by their molecular phenotype. This strategy is based on the general hypothesis that a clinically and morphologically uniform disease may be caused by different molecular mechanisms, that is, that different molecular mechanisms can lead to the same disease or clinical picture. In one of the first studies using this strategy, Perou et al. investigated the gene expression patterns in breast cancer by unsupervised hierarchical cluster analysis. They were able to divide malignant high-grade breast cancer tissue samples into four diverse subgroups with differential gene expression profiles. In several ensuing studies, these subgroups were refined, and their prognostic and therapeutic relevance has been confirmed. Similar studies with impact on the classification of veterinary tumours are lacking but hopefully will be available in the near future.

Hypothesis-driven gene expression profiling is the complete opposite strategy to unsupervised, explorative gene expression profiling. This strategy uses gene expression signatures which either have been identified during in vitro or in vivo experiments using models of the disease of interest or theoretically by comparison with other diseases, species, or consideration of known gene functions. The relevant set of genes or rather "functional" gene expression profiles are applied on preexisting microarray data to subcategorize the tissues according to the activity of the relevant genes. For instance, it has been hypothesized that carcinogenesis is a chronic inflammatory process, and a tumor can therefore be considered as non-healing wound. Based on this idea, the gene expression profiles of serum-activated fibroblasts were applied to the gene expression profiles of different tumours. It has been shown that signature was an independent prognostic factor and showed a strong prediction of metastasis and overall survival and may therefore represent a hypothesis-based prognostic gene set.

Hypothesis-driven gene expression profiling studies on veterinary diseases are not available momentarily, most probably to the delayed accumulation of necessary gene expression data in experimental models. However, due to the successful application of this approach on human diseases, it can be expected that similar studies will be available in the near future.

Metagenomics

Term 'Metagenomics' was coined in the year 1998 by Jo Emily Handelsman an American molecular biologist. She defined the term as "the theoretical collection and analysis of all genomes from members in a microbial community from specific environment". It can also be defined as "Direct genetic analysis of genomes contained within an environmental sample without the prior need for cultivating clonal cultures".

Initially, the term was only used for functional and sequence-based analysis of the collective microbial genomes contained in an environmental sample, but currently it is also widely applied to studies performing polymerase chain reaction (PCR) amplification of certain genes of interest.

Genomics determines the complete genetic complement of an organism by highthroughput sequencing of the base pairs of its DNA. Metagenomics, by contrast, involves sampling the genome sequences of a community of organisms inhabiting a common environment. Metagenomics provides a relatively unbiased view not only of the community structure (species richness and distribution) but also of the functional (metabolic) potential of a community.

One of the indicators that cultured microorganisms did not represent much of the microbial world was the often observed "great plate count anomaly"- the discrepancy between the sizes of populations estimated by dilution plating and by microscopy. The discrepancy has been found to be particularly evident in aquatic environments, in which plate counts and viable cells estimated by acridine orange staining can differ by four to six orders of magnitude, and in soil, in which 0.1 to 1% of bacteria are readily cultivable on culture media under standard growth conditions.

Work on the diversity of soil bacteria, demonstrated with DNA-DNA re-association techniques that the complexity of the bacterial DNA in the soil was multi-fold greater (approx 99.9 fold) than could be accounted for by culturing.

Advances in molecular techniques, when applied to microbial diversity have granted access to a "novel uncultured microbial communities". e.g. Polymerase chain reaction (PCR), rRNA genes cloning and sequencing, Fluorescent in situ hybridization (FISH), Denaturing gradient gel electrophoresis (DGGE & TGGE), Restriction-fragment length polymorphism (RFLP), 2-D Gel Electrophoresis, Next Generation Sequencing (NGS).

Metagenomics Timeline

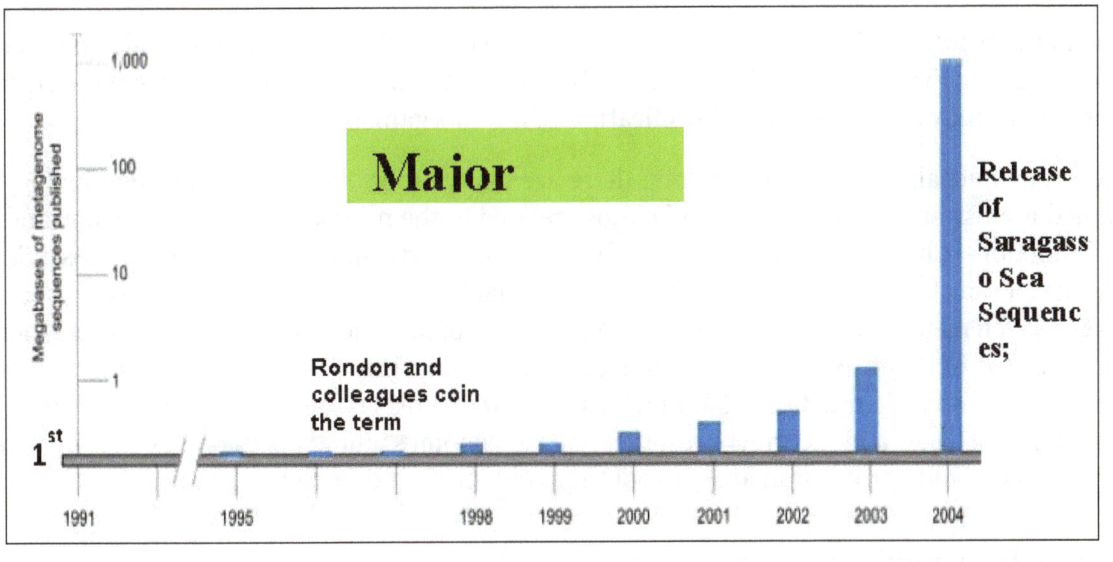

The study of microbial communities has changed from the era of Leeuwenhoek to the era of NGS, including all modern techniques used for the identification and characterization of microbes. Earlier, scientists like Robert Koch tried to isolate the "invisible" organisms by using nutrients in a solid phase, like potato slices or gelatin to cultivate and isolate microorganisms in order to study their morphology microscopically. Later, microscopy became the principal tool for studying microorganisms and their ecological status. Development of practical staining techniques such as Gram, Ziehl-Neelsen, and Schaeffer and Fulton further improved the resolution of microscopy techniques.

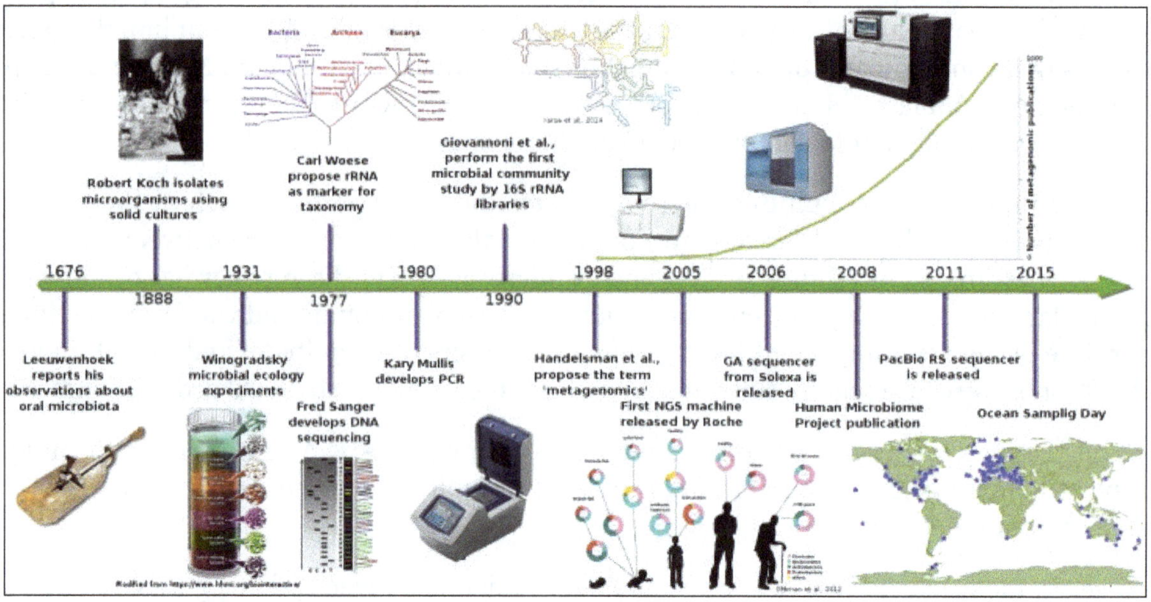

Figure: Metagenomics timeline in the study of community structure and diversity.

Further, Winogradsky's ideas and his contribution to ecology revolutionized microbiology and gave birth to a new concept named "microbial ecology," which refers to the study of microorganisms and their environmental roles. For almost 300 years, the study of microorganisms was based on morphology features, growth, and selection of some biochemical profiles. These techniques provided an insight into the microbial world, but nowadays, they provide only a limited resolution for other applications.

In the late 1970s, Carl Woese proposed the use of ribosomal RNA genes as molecular markers for life classification. This idea in conjunction with the DNA sequencing method revolutionized the phylogenetic study and taxonomic classification of microorganisms.

However, in spite all these improvements; there are many other observations in microbiology that remained unanswered to some extent, like those related to the microorganism's metabolic and ecological functions. Characterization of certain functions in a particular environment was possible only after gene cloning from community DNA of a certain habitat and heterologous expression to associate with a given metabolic function (i.e., enzymatic, key proteins and pathways). Genetic engineering approaches led to the discovery of new molecules and identification of new microbial community members, but with few limitations. Cloning biases, sampling biases, dependence on data bases, misidentification of genes and incorrect promoter sites in genomes, and dispersion of genes involved in secondary metabolite production are some of the problems found in metagenomics.

Reasons for Unculturability

Uncultivable microorganisms might grow in pure culture if provided with all the required chemical components of their natural environment. Interestingly, microorganisms require specific signals originating from their neighbors that indicate the presence of a familiar environment. In the absence of proper signaling, microorganisms will not grow in an unfamiliar environment even in the presence of appropriate nutrients, and this may explain why so many microorganisms cannot be isolated in pure culture on artificial media in vitro.

However, microbes may be recalcitrant to culturing for diverse reasons:

- Lack of necessary symbionts, nutrients, or surfaces.
- Ecological succession, which is not possible in laboratory conditions.
- Excess of inhibitory compounds.
- Incorrect combinations of several physical parameters e.g., temperature, pressure, or atmospheric gas composition.
- Accumulation of toxic waste products from their own cellular metabolism.
- Intrinsically slow growth rate or rapid dispersion from colonies.

Metagenome Technologies

DNA Isolation

Numerous studies have investigated new methods to improve extraction, purification, amplification, and quantification of DNA directly from soil, thus making it possible for a microbiologist to access the unculturable microbial wealth. Comparative studies have been performed to analyze the efficiency of methods for extraction and purification of soil DNA recovered. Humic acids and its derivatives along with polyphenols present in soil are the major contaminants, which inhibit several enzymes of genetic engineering i.e. restriction endonucleases, DNA ligases and DNA polymerase, thus rendering the DNA unfit for further studies.

Many methods have been reported to be useful in eliminating humic acid and polyphenolic contaminants from the environmental sample. There is a long list of treatment methods, which includes addition of polyvinylpolypyrrolidone (PVPP), hexadecyltrimethylammonium bromide (CTAB), divalent metal salts, hydroxyapatite columns, cesium chloride density centrifugations, ion exchange and size exclusion chromatography, silica, activated charcoal, gel electrophoresis etc. Lignin degrading basidiomycetes, collectively referred to as white rot fungi, are common inhabitants of forest litter and fallen trees. These microbes that have been convincingly shown to efficiently depolymerize, degrade, and mineralize soil components e.g. humic acids, humin and fulvic acids. Sharma and co-workers from Delhi University have reported the use of polyphenol oxidase (laccase) from fungal source in soil DNA purification. The nonspecific nature and extraordinary oxidation potential of the peroxidases have attracted considerable interest in the development of several other bio-processes. Moreover, with the development of a DNA isolation system for lignin and humic acid rich environmental sample, Sharma and co-workers has exploited a totally new area and provide a more satisfactory, mechanistic and practical solution to one of the long awaited basic metagenomics problem.

Screening Strategies

Sequence based Screening of Meta-library

Sequence-driven analysis relies on the use of conserved DNA sequences to design hybridization probes or PCR primers to screen metagenomic libraries for clones that contain sequences of interest. Novel genes and pathways have been reported as a result of sequencing of meta-libraries. The conserved sequence/regions of phylogenetically related genome facilitate the isolation of genes without prior knowledge of the full sequence. By contrast, the sequences of most genes of practical importance are far too divergent to make it possible to identify new homologues by PCR or hybridization.

High-throughput sequencing technologies have recently popularized shotgun metagenomic and 16S ribosomal DNA sequencing methods, which identify members and/or functional genes at a greater scale and in more detail. Their use in diverse environments has revealed the presence of new division of archaea (Korarchaeota), symbiotic relationships between gut microbes and human health and characterized the nutrition systems involved in symbiosis. Even more applications of these methods are used in food science, pharmaceuticals and forensic science. Many large-scale metagenomic projects are now generating comprehensive microbial sequence collections for different environments (e.g., human-associated, soil, and ocean environments).

Large-scale sequencing projects such as the one initiated by Craig Venter for the metagenome of the Sargasso Sea resulted in the identification of numerous novel genes and is a very famous example of sequence-based metagenome analyses. Similar approaches have been initiated by European laboratories to sequence complete or partial metagenomes of a phylogenetically highly diverse biofilm. Complementary to this, Tyson and coworkers described the nearly complete sequence analyses of the metagenome of an acidophilic biofilm, which was characterized by a low biodiversity.

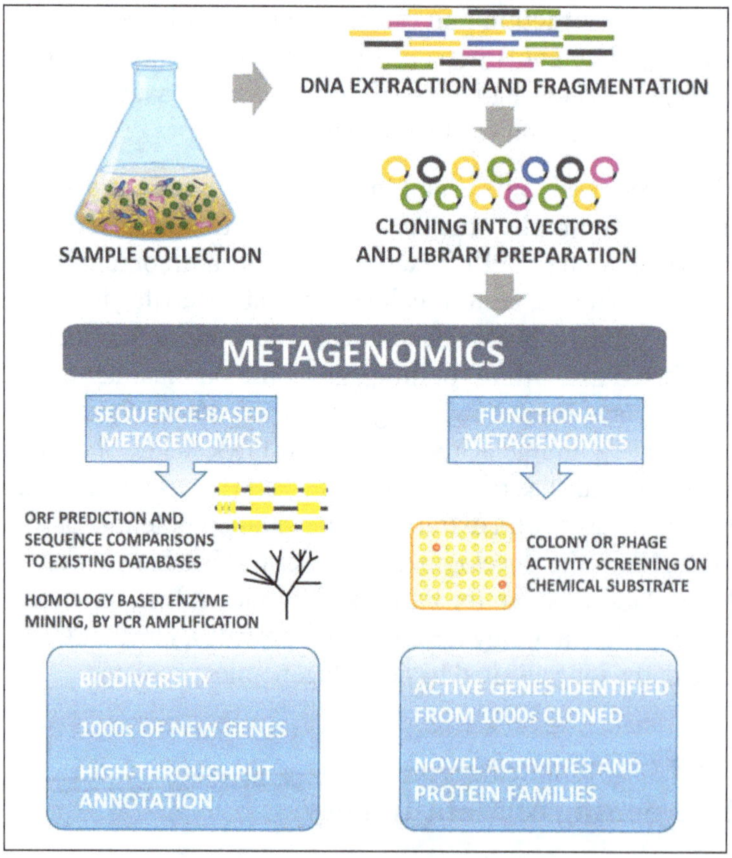

Figure: Functional and sequence driven screening of metagenome library.

Functional Metagenomics Analysis (Functional Annotation)

The function-driven analysis is initiated by expression of clones processing their native promoter on selected substrate followed by characterization of the positive clones by gene sequencing and biochemical characterization. This approach quickly identifies clones that have potential applications in medicine, agriculture or industry working on enzyme based products. The limitations of the approach are that it requires expression of the function of interest in the host cell and clustering of all of the genes required for the function. Functional metagenomics also depends on the availability of screening technologies that can be performed efficiently on several meta-libraries. Many approaches are being developed to mitigate these limitations. Improved systems for heterologous gene expression are being developed with shuttle vectors that facilitate screening of the metagenomic DNA. Several modifications in Escherichia coli host have further expanded the gene expression from complex natural environment. Functional screens of metagenomic libraries have identified novel antibiotic resistance gene, industrial enzymes (e.g., lipases, cellulases, xylanases, laccases, chitinases etc.), membrane proteins, genes encoding several metabolic pathways.

Gut Metagenome

Insect Gut

Microbes colonize the insect gut through food and play a significant role in digestion and metabolism of complex substrates. The complexity of the insect digestive tract, unculturable microbes, and the lack of genetic tools for most cultivable microbes make it challenging to explore the nature of microbial interactions within an ecological niche. Comprehensive analysis of the bacterial diversity within a host species is a prerequisite in both insect physiology and microbial ecology to allow a better understanding of the ecological roles of insect symbionts and interactions with their insect hosts. However, most studies on gut microbial diversity have been taxon specific, focusing on insects such as termites, ants, fire bugs, fruit flies, beetles, and bees, which can be used to study the diversity function relation across all insects. The insect gut microbiotas are dominated by Proteobacteria (62.1% of the total reads, including 14.1% Wolbachia sequences) and Firmicutes (20.7%). Significant differences are reported in the relative abundances of anaerobes in insects and are classified according to the criteria of host environmental habitat, diet, developmental stage, and phylogeny. Gut bacterial diversity is significantly higher in omnivorous insects than in stenophagous (carnivorous and herbivorous) insects.

Metagenomic analysis of the leech crop (digestive tract) provides insight into gene content and possible metabolic functions that the leech gut symbionts possess and might contribute to the host. Studies on the microbial composition of the crop using both culture-dependent and - independent approaches have suggested a very simple microbiome in the leech crop that is dominated by two species: Aeromonas veronii and a Rikenella-like bacterium. A. veronii is a Gram-negative, facultative anaerobe that has been shown to lyse erythrocytes present in the crop and is resistant to hemocyte phagocytosis.

Blood is poor in vitamins; therefore, microbial production of vitamins is likely to be critical for leech health. Within the metagenome of crop, genes are found that are likely important for survival of the symbiont and contribution to the host. Insects with restrictive diets often possess symbionts that produce essential amino acids and vitamins missing from the host diet. Blood suckers and feeders (e.g., mosquitos, leech and bed bugs), harbor microbial symbionts that produce essential vitamins and amino acids.

Globally, termites are an extremely successful group of wood-degrading organisms and are therefore important both for their roles in carbon turnover in the environment and as potential sources of biochemical catalysts for efforts aimed at converting wood into biofuels. In the year 2008, Warnecke and co-workers have reported any direct role for the symbiotic bacteria in the gut of the termite in cellulose and xylan hydrolysis. Here we use a metagenomic analysis of the bacterial community resident in the hindgut paunch of a wood-feeding Nasutitermes species to show the presence of a large, diverse set of bacterial genes for cellulose and xylan hydrolysis.

Termite guts are characterized by steep physical, chemical and biological gradients. Human colonic and bovine ruminal microbial communities are dominated by non-motile microbes, but termite hindguts have long been noted as sources of highly motile bacteria. Motility in bacteria is often coupled with chemotactic behavior for which about 1,500 genes are known. Termite gut metagenome study illustrates complex environment with rich reservoirs of bacterial enzymes relevant both to reaching a better understanding of the fundamental process of wood degradation, and to engineering novel schemes for the conversion of lignocellulose into biofuels and other microbial products of interest. New insights into other important symbiotic functions including H_2 metabolism, CO_2-reductive acetogenesis and N_2 fixation are also provided by gene analysis of a termite gut microbial community.

The arthropod gut is a differentiated organ harboring a complex biotope comprising both resident and transient members from protozoal, bacterial, and archaeal genera. Many of these organisms are symbionts that contribute in a concerted way to a complex chemical cycle sustaining the metabolic competence of the host. These symbiotic relationships help to define the metabolic traits of the insect, contributing to efficient sharing of the derived nutrients. In addition to carbon.

Metabolism, meta-data have demonstrated that the microbial consortia contribute to nitrogen cycling, acetogenesis and methanogenesis, and prevention of foreign microbial pathogenesis. Metagenomics have resulted in the discovery of thousands of new enzymes with novel sequences and often with novel activities, stabilities, and operational optima. Direct DNA extraction from arthropod gut has been used to target the discovery of novel xylanases with a purpose to understand their evolutionary relationships to other well characterize xylanases and to explore their potential in hemicellulose degradation.

Human Gut

The human intestinal microbiota is composed of 10^{13} to 10^{14} microorganisms, which together ("microbiome") contribute ~100 times as many genes as our own genome. Gill and coworkers in the year 2006 analyzed human distal gut microbiome and found ~78 million base pairs of unique DNA sequence. Using metabolic function they analyzed the identified genes and compared human genome with the average content of previously sequenced microbial genomes. Our gut microbiome has significantly enriched metabolism of glycans, amino acids, and xenobiotics; methanogenesis; and 2-methyl- D -erythritol 4-phosphate pathway–mediated biosynthesis of vitamins and isoprenoids.

Progress made with 16S rDNA-based studies has reveled significant differences in community membership between healthy adults, differences that may contribute to variations in normal physiology between individuals or that may predispose to disease. For example, studies of humans and gnotobiotic mouse models indicate that mutualistic relations with the gut microbiota influence maturation of the immune system, modulate responses to epithelial cell injury, affect energy balance, and support biotransformations that we are ill equipped to perform on our own, including processing of xenobiotics. As microbial genomes contribute significant pathways for key metabolism thus, we can say that humans are super organisms due to combined attributes of both microbes and human.

Metaproteomics

The recent availability of extensive metagenomic sequences from various environmental microbial communities has extended the postgenomic era to the field of environmental microbiology. Although still restricted to a small number of studies, metaproteomic investigations have revealed interesting aspects of functional gene expression within microbial habitats that contain limited microbial diversity. These studies highlight the potential of proteomics for the study of microbial consortia. Metaproteomic investigations of environmental samples will be most potent when coupled with information about species diversity and richness within the ecosystem. Thus, ecological niches with simpler microbial communities [e.g. acid mine drainage (AMD) biofilm, whale-fall sites and activated sludge] should be easier to understand than more-complex ecosystems (e.g. forest, ocean and soil) from their proteomic signatures.

Metaproteomics has been used to decipher the metabolic details of a treatment process for activated sludge wastewater that operates for phosphorus removal. The process, known as enhanced biological phosphorus removal (EBPR), is desirable because it enables the microbiological removal of phosphate from refuse water and, thus, limits eutrophication in surface waters that receive effluent. The treatment process is characterized by alternating anaerobic and aerobic phases, with most of the sludge being recycled. This arrangement selects for specific polyphosphate-accumulating organisms (PAOs) such as uncultured organisms that are related to Rhodocyclus species.

Biotechnological Prospects

Metagenomics has identified a significant number of novel genes encoding for fine chemicals, biocatalysts with high potential for use in pharmaceutical products or production processes, enzyme-catalyzed chiral synthesis. Interestingly, searches in metagenome-derived DNA libraries have mainly focused on a rather small group of enzymes. Among these are lipases and esterases. Lipases and esterases are of importance for biotechnological applications because many enzymes of this class remain active in organic solvents; they usually display exquisite chemo-, regio-, and stereoslectivities and they do not require cofactors. Similarly, polysaccharide-modifying enzymes such as the starch modifying enzymes are of considerable interest to industry. Therefore, a significant number of metagenome searches have identified polysaccharide modifying genes. Further, the isolation of enzymes useful for the production of bulk chemicals, proteases and nitrilases has been reported. Metagenome searches have also focused on the isolation of genes involved in vitamin biosynthesis.

Animal Forensics and Applications

Animal forensics is defined as the scientific techniques used in the investigation to solve problem of crimes against animals such as agricultural, domestic and wild animals. Animals forensic has wide range of applications. It helps in preventing illegal trafficking of regulated species in the case of their body parts along with products, in solving cases linked with the crime against animals, legal claims settling associated to endangered animals identifying due to unknown death causes, protected and threatened animals individuals etc. Animal's forensics help in protecting religious and social values of a society because there are many religious issues related with the diets and animals also used as a part of human diet. In addition to these there are many challenges to animal forensics as well. These include low availability of very old as well as degraded samples, specimens obtained from closely associated species etc. Over the past 20 years, the analysis of DNA is one particular biological tool has solved all the problems and revolutionized the forensic studies by analysis of DNA. Hence, all living beings possess DNA and exhibits variability both among as well as within species, DNA investigations has associated with a legal case contains in it information relating to the source. DNA analysis has now become an indispensable technique of current forensic science consisting animal forensics, using extremely sensitive and specific PCR based methods in analyzing biological materials. Evidences can be associated to crime clues by DNA evidence from as a low amount of saliva present on a cigarette butt as well as on skin cells of a steering wheel. Hence case can be solved after analyzing the DNA present in the suspected clues. This chapter mainly emphasis on animal forensic using the DNA analysis for describing the application of animal forensics.

Material used in Forensics

Hair

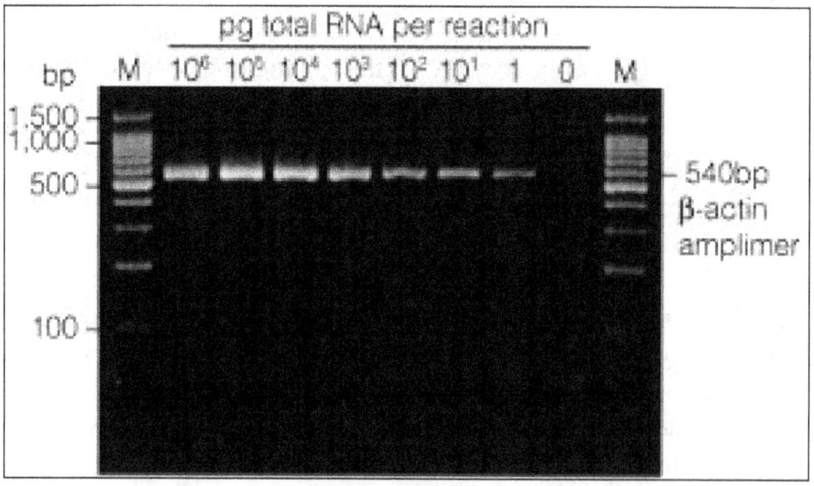

PCR amplified

Comparative morphology, histology along with microscopy shows the classical tools in the stream of forensic analysis and hair is used as trace evidence. Every mammalian species has hair with specific feature of colour, length and root structure as well as with various morphological characteristics. Hairs possess extremely low amounts of DNA. Because nuclear DNA content is too low from hair which is not always successfully amplified by PCR, showing the presence of lots of PCR inhibitors in the samples extracted. Hair pigment melanin one of the strong inhibitor present in the PCR process. More prominently, hair-dying was also has a strong influence during PCR process. Different DNA extraction methods of were used but Chelex method was most recommended for PCR amplification method. This method is cost effective, eco-friendly, quick, simpler and can also be used for DNA extraction from skin, blood, bone, muscle tissue, feather, faecal sample etc.

Chelex method is eco-friendly, cost effective, simpler, quick and easy to perform, and can also be used for DNA extraction from blood, skin, feather, bone, muscle tissue, faecal sample, etc.

Blood

The determination of serum proteins, blood group serology, and isoenzymes, along with MHCantigens characterization are the classical analytical methods of blood for blood analysis. But currently far blood analysis possibilities whole spectrum blood molecular methods are used. The DNA is extracted from the white blood nucleated cells is used for DNA extraction. Appropriate method choice for analysis depends on quality and quantity of the sample available. Due to its convenience and ease in dried blood samples collection and transportation, it becomes very popular among forensic as well as other public health uses. It has been found that biological markers associated in it, can be measured from serum/plasma and whole blood can be investigated from dried blood clues. Nguyen and his co-workers investigated a two-step lysis method for dried blood samples genomic DNA extraction. The method was based on the Sumota Chaisomchit one step lysis extraction method. Two step methods was found to be more efficient as well as cost effective when compared with one step method and QIAmp DNA Mini Kit based commercial method.

Saliva

In forensic sciences in addition to whole blood and hairs, saliva and saliva stains, can use as an alternative source for DNA standards. For DNA isolated from saliva various materials can be used such as buccal swabs, envelops, and cigarettes. For extraction of DNA from buccal swabbing, cotton is removing from the swab stem and transfer in to a microfuge tube. Following by addition of lysis buffer along with Proteinase K (20mg/ml). To extract DNA from the cigarettes, we can remove the cigarettes filter ends and cut into pieces, and lyses with lysis buffer and Proteinase K in a microcentrifuge tube. Remaining steps of the protocol is similar for all the sample types. Incubation of the sample for overnight at 56 oC. Finally the DNA is extracted with an equal volume of phenol/chloroform. Additon of glycogen increase the precipitation of the DNA with isopropanol.

Bone and Teeth

The protein mineral matrix present in bone and teeth samples have an effective physical and chemical barrier to biological attack and environmental deterioration and used only and almost the best, biological material for DNA typing.

Feathers

The use of feathers for extraction of genomic DNA from avain, to know the bird age and size. For DNA extraction from features, nonivasive method has been used.

Skin and Leather

Once tanned it becomes very difficult to discriminate skin/leather of different species of wild animal from domestic animal. By advantage of this only, commercial products from wild animals are usually sold as if they from domestic animals and vice versa. In these cases, only DNA analysis is solution for differentiating between different species processed skin samples.

Applications of Animal Forensics

Applications of animal forensics are described below:

- Animals can be the Victim: In the Argentina case report ranch cattle were stolen as well as slaughtered. After that it could be found uniquely due to its brand. Their DNA profiles were found different when compared with meat chunks from butcher's shop and it could identify as killed cattle. There are many such type of cases of animals cruelty which include bestiality, animals killing for skin, bone etc. In these cases forensic scientist is expected to identify and link the traces of animal tissues available from the crime scene to the animals being killed and solved the cases.
- Animals can be the Perpetrator: Animals can be acts as well perpetrators as victims and involved in an attack animal. Thus identification of such type of animals is required because such animal causing property damage and accident. Thus again forensic scientist is helpful to link clues of animal from material obtained from the crime scene to identify the animal involved in crime.
- Wildlife Forensics: Illegal trade of animals as well as their products is growing in global black market. Since weight-for weight wildlife trade is more profitable as compared to drugs, and associated with less risk and wildlife trade is attractive for criminals. It creates serious harms, to both indirect and direct, along with the global biodiversity. During this process a series of steps occur that leads to s to the species involved extinction. The species involved for trade market are directly affected due to over exploitation. Wild populations over exploitation can cause its global extinction. Wildlife trade also gives a way for the introduction of outside species infected with various diseases. These are the source of infection in native species and then disease become invasive among native species. A special field of animal forensics is known as wildlife DNA forensics that deals with such cases. The ultimate objective of wildlife DNA forensics is identification of clues in order to know the population, species, relationship or individual identity from clue/sample.

Forensic Genetic Identification Methods

Molecular animal forensics involves various genetic tools for identification and characterisation of a sample from crime scene or involved in illegal wildlife traders as well as in black markets. The genetic identification of animals can be done at three different levels described below:

Species Identification

This is the most common used application in wildlife DNA forensics and is lower common in human forensic for identification the species from available clue using genetic analysis. This method is used in species identification in cases illegal poaching form a suspect's possessions. It has specific use in the investigation of traded products that is lost their morphological characters.

Genetic species identification is associated with the isolation and analysis of different DNA markers found in eukaryotic genome so that have species specific variation. There are two types of genetic markers are used for species identification i.e. mitochondrial and nuclear markers.

Mitochondrial Markers

There are 12srDNA and 16s rDNA, two ribosomal RNA genes, present with in animal mitochondria. From these highly conserved is 12s rDNA and used in identify the higher species. Conserved primers are used in molecular analysis, to amplify these markers of identification and further amplicons are sequenced. After that sequencing data are compared by the alignment of these sequences with reference sequences using specific bioinformatics tools.

Nuclear Markers

Different nuclear markers has been used for DNA fingerprinting include amplified fragment length polymorphism (AFLP), random amplified polymorphic DNA, and micro satellites. However, the generally used markers are mitochondrial DNA markers in animals specifically cytochrome b and cytochrome oxidase subunit 1. Moreover, where mitochondrial DNA is present in multiple copies in comparison to nuclear DNA, which present only in 2 copies of each marker in a species. Hence the success chances increase significantly during analyzing evidence samples that typically contain relatively lower cellular material.

DNA Nucleotide Sequencing

It is a primary and most important method for species identification, and generally involves sequencing of around 500 bases of DNA to give a species specific sequence. Sequencing identifies every nucleotide within a specific target region of DNA, this method have specific advantages because it involves the universal PCR primers designing that can be used for DNA amplifications from wide range of species without any previous knowledge associated to the sample. It also involves species specific primers designing, hence in a single multiplex PCR reaction; more than one species can be identified.

In this method various suspected samples are processed for DNA isolation. After that they are amplified using the species specific primers and proceed for sequencing and got more than 500 bps sequences. Their identification is done the use of reference having similarities of 90 to 95% or higher from related species using appropriate bioinformatics tools. From theses sequences we can generate at a phylogenetic tree, and find the evolutionary relationships among the suspected samples and reference sequences from matched species. The laying position of suspected sample in tree gives the nearest reference species to be identified as the suspected source.

SNP Typing or Genotyping

The differences identified among species at genetic markers are mainly due to single base pair variation in the every species DNA sequence, called as single nucleotide polymorphisms (SNP).

Thus, SNP typing/genotyping provides specific variable region in a DNA fragment to be identified allowing species differentiation via faster as well as cheaper tests that are not require such long segments of good quality DNA as required at the time of DNA sequencing techniques. However, lesser information is obtained in case of SNP typing in comparison to DNA sequencing. SNP typing needs smaller DNA sequences and thus, the size of the genetic marker targeted is reduces. Hence, there is increased risk of misidentification during using SNP typing.

Random Amplified Polymorphic DNA (RAPD)

It is a simple and least expensive technique. During this technique arbitrary primers of usually 10 bases long are used for amplification of random segments DNA. No previous information is required about the DNA sequence related to the targeted gene because the primers will bind anywhere in the genome sequence. If the targeted DNA encompasses sequences complimentary to the primer nucleotide, amplified PCR products will be detected otherwise if it does not give PCR products if there are no complimentary sequences to the primers. Therefore, in this method we have tried different random primers until we found a pair of random primers that provide us a unique DNA profile for each targeted DNA sample. This technique relies only on an intact, large DNA sequence and cannot be used at the time of degraded DNA samples.

In other cases, bands specific to the target species are found from the DNA fingerprint are identified, extracted and sequenced. After that these sequences are used to design primers that will specifically provide the amplified product from the species specific region, known as a sequence characterized amplified region (SCAR). Further, these primers sequences can be used for quick species identification. This technique is a highly reproducible.

Amplified Fragment Length Polymorphism

This method combines features of both RFLP and RAPLD analysis and having greater differentiation power as compared to RAPD. This technique is a PCR based method, and uses restriction enzyme digested fragment of template during PCR amplification using the adaptor sequence complementary primers, the restriction site sequence, and a fewer nucleotides from inside the restriction site fragments. The amplified products are then visualized using denaturing polyacrylamide gels.

Micro Satellite Genotyping

This technique is more reproducible than SNP genotyping, RAPD and AFLP analysis because of micro satellite DNA markers hyper-variability. For the species identification purpose only those micro satellite markers are selected that have a significant variation among species but are commonly conserved within species. This method is best tool for parentage analysis. These are also known as hypervariable markers and short tandem repeats as well as simple sequence repeats. They are short sequence motifs contains two to four nucleotides and it can be repeated for three to 100 times. One general example is (CA)n repeat, here n means any base pairs(bps) in place of n, that is most commonly found in human as well as other genomes, are present each few thousands bps.

Microsatellites are in the form of bi-allelic/multi-allelic form in a population and codomimnant markers. Theses alleles are transferred from one generation to the next generation. These alleles are visualized on polyacrylamide gels after their amplification by PCR. Only those individuals give same bands that are homogygote; however heterozygote gives different sizes band. Hence microsatellite is very important during progenitor identification of individuals as well as suitable for peternity determining.

DNA profiles are formed from different samples during analysis for individual identification. If different DNA profiles produce from two samples, then they did not originate from same individual. However if same bands are produced from two different samples it surely originated from same individual. The genetic band patterns can be produced using the information from microsatellite loci/sex determination markers and from the mitochondrial control region.

There are many situations when animals are sold for their meat as well as their high processed meat products in case of wild life forensics. Individual identification on the basis of their specific genetic profile has an important application to monitor the different types of animals entering for commercial marketing. Similarly, in order to differentiate the legally as well as illegally obtained samples, individual identification is required.

2

Animal Cell Culture

The process of growing animal cells in vitro on a favorable artificial medium and in a favorable environment is known as animal cell culture. There are various types of animal cell culture such as primary and secondary cell culture, and subculture of cell. This chapter explains all these types of animal cell cultures and various other aspects related to it such as organ culture, cell cryopreservation, etc.

Animal cell culture involves the in vitro (in the laboratory) maintenance, propagation and cultivation of animal cells. These cells are previously growing in a human or animal which will be are modified to cultivate in plastic or glass vessels in a closed sterile environment which mimics our body condition in which cells were previously growing. In animal cell culture, cells are kept in a specialized CO_2 incubator to stay at body temperature of 37 °C and proper humidity. Cells are provided with special nutrient rich culture media so that these cells can grow and divide. Utmost care should be taken to maintain aseptic environment so that cell will grow free of any microorganism. For same purpose sometimes antibiotics and antimycotics are also added to the culture medium.

Requirements for Animal Cell Culture

Media and Reagents

Many cell culture medium and reagents are required in maintaining and propagation of Animal cell culture. All these medium and reagents should be in appropriate amount and in sterile from authentic source.

- Media: The cell culture medium is principal component of the culture. Medium provide the necessary nutrients, growth factors & hormones for cell culture and growth. Medium is also responsible for regulating pH and osmotic pressure of culture. Historically natural medium from tissue extracts of organism body fluids was utilized but these need to be sterilized. Due to development in technology and increase in the demand of these medium, now a day available in powder form or ready to use. Basically, three types of media available: a) basal media b) reduced serum media c) serum-free media are available from different authentic supplier. The medium should be select according to the nature of cell-line. Some cell-line in a specific medium while some cell-line can be grows in any medium or broad range of medium.
- Serum: Animal cell is culture is very complex procedure. Serum is very helpful in growth and adhesion as well as provides hormones, lipids and minerals to cells. Generally, 10% serum provide in basal media for cell culture. Serum should be purchase from source which provides the guaranteed for their quality, safety and consistency.

Types of Media

- Basal Media: Medium should be selected according to the nature of cell-line. Although most celllines grow well in basal media like MEM, DMEM, RPI etc. Basal medium should have amino acids, vitamins, inorganic salts and source of energy in appropriate amount. Basal medium could all components therefore supplemented with 10% serum.
- Reduced-Serum Media: Reduce-serum medium can be utilized to decrease the undesired effects of serum in cell culture experiments. Reduced-serum media contains the basal media formulations enriched with nutrients and animal-derived factors, which reduce amount of serum that is needed.
- Serum-Free Media: Although serum has many advantage for cell culture and could not found any alternate of serum. There are many disadvantage of serum also like high cost, standardization, specificity, variability and interfere with cellular growth. Serum helpful for consistent performance, easier purification, precise evaluation of cellular functions, increased productivity and better control over physiological response. But there are certain disadvantages of serum free medium like cell type-specific media formulations require, highly purify reagent and slow growth.

Basic Requirement for Culture Medium

Nutritional Components

Cells need the basic nutritional conditions to grow in vitro, including:

- Amino Acid: Amino acid is the raw material for the cell to synthesize protein. All the cells need twelve essential amino-acids: arginine, cystine, isoleucine, leucine, lysine, methionine, phenylalanine, threonine, tryptophan, histidine, tyrosine and valine, which are L-amino acids. Furthermore, glutamine is another component playing important role in the cell metabolism process. The nitrogen contained in glutamine is not only the source of purine and pyrimidine of nucleic acid, but also the essential material for the synthesis of the Tri-, bi-, mono-phosphate acid glycosides.
- Monosaccharide: Cultured cells use aerobic glycolysis and anaerobic glycolysis of hexose as main energy source. In addition, hexose is used for the synthesis of some amino acid, fat and nucleic acid. Cell absorptive capacity varies among different monosaccharides, with the highest for glucose and the lowest for galactose.
- Vitamin: Vitamins mainly act as coenzymes or prothetic groups in cell metabolism processes. Biotin, folate, nicotinamide, pantothenic acid, pyridoxine, riboflavin, thiamine and vitamin B12 are common component in culture medium.
- Inorganic Ion and Trace Element: Besides some basic elements (including sodium, potassium, calcium, magnesium, nitrogen and phosphorus), cell growth needs some trace elements, such as molybdenum, vanadium, iron, zinc and selenium, copper, manganese.

Somatomedin and Hormones

Cells grown *in vivo* are always regulated by somatomedin and hormones. Many researches demonstrate that various somatomedin and hormones are very important to maintain cell function and status (differentiated or undifferentiated). Some hormones have promoting growth effects on different cell types. For instance, insulin can promote the use of glucose and amino acids in the cell. Some hormones are cell-type specific, as hydrocortisone that can promote the growth of epidermal cells and prolaction that induces the proliferation of mammary epithelial cell.

Osmotic Pressure

Cells need an isotonic environment and human plasma osmotic pressure is about 290 mOsm/kg, which is thought to be ideal osmotic pressure to culture human cells. Mouse plasma osmotic pressure is about 320 mOsm/kg. Osmotic pressure of 260-320 mOsm/kg fits for most mammalian cells.

pH

The suitable pH for most cells is 7.2-7.4; otherwise it will produce harmful effects. The culture medium should have some buffer capacity. The main substance causing pH changes is CO_2 produced in cell metabolism process. In an airtight environment, CO_2 can combine H_2O_2 to produce carbonic acid and thus reduce the pH value of the medium. Synthesized medium employs $NaHCO_3$-CO_2 buffer system to solve this problem. In the buffer system, the boost in [H^+] increases the reaction rate H^+ + salt => weak acid and takes some H^+ out of circulation. It is based on the constant equilibrium.

$$\text{NaHCO3} + H_2O \longleftrightarrow Na^+ + HO^- + H_2O + CO_2 \uparrow$$

Natural Medium

Natural medium is described as animal body fluids or medium of tissue extraction, including plasma, serum, lymph, chicken embryos leaching solution. Natural medium contains rich nutrients, various somatomedin and hormones, similar osmotic pressure and pH to body environment. As this medium has a very complicated production process and big batch-to-batch variation, the medium is gradually replaced by the synthetic medium. Today, serum is still the widely used natural medium.

Serum

Types of Serum

Serum can derive from different animals. Current serum used in tissue culture is cattle serum. Human serum, horse serum is used for some specific cells. Cattle serum has several advantages as used in cell culture: adequate resource, mature preparation technique, long application time.

Cattle serum includes bovine calf serum, newborn calf serum and fetal bovine serum. Take sample for cattle serum, fetal bovine serum derives from caesarean section fetal bovine; newborn calf serum comes from newborn calf born within 24h; bovine calf serum comes from calf with 10 to 30 days. Fetal bovine serum has highest quality because the fetal bovine doesn't expose to outside environment and has lowest antibodies and complement.

Main Components of Serum and its Function

Serum is made from plasma by removing hemaleucin and contains various plasma protein, polypeptide, fat, carbohydrate, growth factor, hormones and inorganic mineral, etc. All these substances keep the physiological balance of promoting or inhibiting cell growth. The following table shows the main components of serum and their mean concentration.

Table: The main components of serum and their mean concentration.

Component	Mean concentration	Component	Mean concentration
Na^+	137mol/L	Alkaline phosphomonoesterase	225U/L
K^+	11 mol/L	Lactic dehydrogenase	860U/L
Cl^-	103 mol/L	Insulin	0.4μg/L
SeO_3^{2-}	26μg/L	Thyroid stimulator	1.2μg/L
Ca^{2+}	136mg/L	Folliclestimulating hormone	9.5μg/L
Fibonectin	35 mg/L	Bovine somatotropin	39μg/L
Urea acid	29 mg/L	Prolactin	17μg/L
Creatine	31mg/L	T_3	1.2μg/L
Hemoglobin	113 mg/L	Cholesterol	310μg/L
Bilirubin(total)	4 mg/L	Cortisone	0.5μg/L
Inorganic phosphorus	100mg/L	Testosterone	0.4μg/L
Glucose	1250mg/L	Progesterone	80μg/L
Urea	160mg/L	Prostaglandin E	6μg/L
Total protein	38g/L	Prostaglandin F	12μg/L
Albumin	23g/L	Vitamin A	90μg/L
α_2- macroglobulin	3g/L	Vitamin E	1 mg/L
Endotoxin	0.35μg/L	Fe^{2+},Zn^{2+}, Cu^{2+},Mn^{2+},Co^{2+}, Co^{3-}, etc.	μg/L to ng/L

The main function of serum is listed as follow:

- Provide Essential Nutrients: Serum contains various amino acids, vitamins, inorganic min erals, fat, and nucleic acid derivatives, which are essential nutrients for cell growth.
- Provide Adherence and Extension Factor: Many cells cultured in vitro have to attach the culture vessel to grow, which is dependent on extracellular matrix. Cells can secrete extracellular matrix in vivo, but this ability will decrease or even disappear according to the increment of passages. Serum contains some components, (fibronection, laminin, etc.), which can promote cell adherence.
- Provide Hormone and Various Growth Factors: Serum contains various hormones, such as insulin, adrenocortical hormone (hydrocortisone, dexamethasone), steroid hormone (estradiol, testosterone, and progesterone), etc. The growth factors include fibroblast growth factor (FGF), epidermal growth factor (EGF), peteletdericed growth factor (PDGF), and others.
- Provide Binding Proteins: The binding proteins carry low molecular weight material. For example, the albumin carries vitamins, fat (fatty acid, cholesterol) and hormones. Transferrin carries iron.
- Provide Protection for Some Specific Cells: Some cells (such as epithelial cells, myeloid

cells) can release protease, which can be neutralized by the anti-protease ingredient in the serum. Serum is widely used to terminate the effect of the trypsin. Serum albumin facilitates the serum viscosity and protects the cell from mechanical damage, especially in the suspension cell culture. The trace elements and ions, such as SeO_3 and Selenium, play very important role in metabolic detoxification.

Disadvantages of using Serum in Tissue Culture

The composition of serum is complicated, including favorable components and unavoidable harmful ingredient. The disadvantages of using serum in tissue culture are listed as follow:

- For most cells, serum is not physiological fluid in vivo. Cells only contact serum during the injury healing or blood coagulation. Thus the utilization of serum may change the normal condition of some cell. Serum may promote the growth of some cell (such as fibroblast) and inhibit the proliferation of other cells (such as epidermal kerotinocyte).
- Some components may be toxic to cells. Take polyamine oxidase for example, it can react with the polyamine (such as spermine and spermidine) to form the toxic poly-spermine in highly proliferated cells. Furthermore, complements, antibodies and bacteriotoxin can affect the cell growth, or even lead to cell death.
- Each batch of serum varies from others and the component cannot maintain uniformity.
- The productive process may infect with mycoplasma or virus, which potentially affect cells and lead to the fail experiment and unreliable experimental results.

Rat Tail Collagen

Rat tail collagen, used as either a thin layer on tissue-culture surfaces to enhance cell attachment and proliferation, or as a gel to promote expression of cell-specific morphology and function. This product is ideal for coating of surfaces, providing preparation of thin layers for culturing cells, or use as a solid gel. Rat Tail Collagen is suitable for applications using a variety of cell lines including hepatocytes, fibroblasts and epithelial cells.

Synthetic Medium

The synthetic medium is artificial designed and prepared medium. Nowadays the synthetic medium already becomes standardized commodity with wide varieties and convenience to use.

Basic Medium

Basic Components

After experimental selection, the simplest medium is minimum essential media (MEM), which contains more than 20 kinds of substance and can be divided into 4 subgroups:

- Inorganic salt: $CaCl_2$, KCl, $MgSO_4$, NaCl, $NaHCO_3$, NaH_2PO_4.
- Amino acid: Arginine, cystine, isoleucine, leucine, lysine, methionine, phenylalanine, threonine, tryptophan, histidine, tyrosine and valine.

- Vitamins: Partial polyoxometalates calcium, choline chloride, folic acid, inositol, nicotinamide, pyridoxine, riboflavin, thiamine.

Carbohydrate: Glucose

Besides the above substance related to cell growth, the medium usually also uses phenol red as pH indicator.

MEM is not common basic medium and common medium has more than 30 kinds of component, such as RPMI1640, DMEM. These mediums generally add some non-essential amino acids and vitamins, including serine, proline, biotin, Vitamin B12, etc.

Category

Nowadays, there are more than tens of basic medium available. The most common ones are listed as follow:

- MEM: MEM, also called Eagle's minimal essential medium, is a cell culture medium developed by Harry Eagle that can be used to maintain cells in tissue culture. It only contains 12 kinds of non-essential animo acids, glutamine, 8 vitamins and some basic Inorganic salts.
- DMEM: A variation of MEM, called Dulbecco's modified Eagle's medium (DMEM), (Dulbecco/Vogt modified Eagle's minimal essential medium), contains approximately four times as much of the vitamins and amino acids present in the original formula and two to four times as much glucose. Additionally, it contains iron and phenol red. DMEM is further divide into high-glucose type (4500g/L glucose) and low-glucose type (1000g/L glucose). High-glucose DMEM is suitable for some tumor cells with faster growth speed and difficult attachment, as it is beneficial to retain and grow in one place.
- IMDM: It is modified by Iscove basic DMEM and contains 42 ingredients. It includes selenium as well as additional amino acids and vitamins. In addition, this unique medium lacks iron, with potassium nitrate replacing ferric nitrate. It is well suited for difficult proliferating, low-density cell cultures, including hybrid cell selection after cell fusion, selection of transformed cell after DNA transfection.
- RPMI1640: Roswell Park Memorial Institute medium, commonly referred to as RPMI, is a form of medium used in cell culture and tissue culture. The initial formula is suitable for growth of the suspension cells, mainly for lymphoid cells. This medium contains a great deal of phosphate and is formulated for use in a 5% carbon dioxide atmosphere. RPMI1640, the most mature improved medium, is suitable for most types of cells, including tumor cells, normal cells, primary culture cells, passage cell. RPMI1640 is one of the most common used medium.
- 199/109 Medium: 199 medium is developed by Morgan and his coworkers in 1950 and is one of the earliest culture medium. It was originally developed as a completely defined media formulation for chick embryo cell culture. 199 medium has more than 60 components and contain almost all the amino acids, vitamins, growth hormone, nucleic acid derivative, etc. 109 medium is improved based on 199 medium and better formulated for the cell culture in a serum-free environment.

- Ham F-10/Ham F-12: Ham's F-10 medium is a classical media designed by Ham to support the growth of mouse and human diploid cells in 1962. Ham's F-12, as improved products, has been used for the growth of primary rat hepatocytes and rat prostate epithelial cells. A clonal toxicity assay using CHO cells has also been reported with Ham's F-12 as the medium of choice. Ham's F-12 is also available with 25 mM HEPES buffer that provides more effective buffering in the optimum pH range of 7.2-7.4.
- Serum-free Medium: Serum free media (SFM) are important tools that allow researchers to grow a specific cell type or perform a specific application in the absence of serum.

Advantages of using serum free media include:

- Easier purification and downstream processing.
- Precise evaluations of cellular function.
- Increased definition.
- More consistent performance.
- Increased growth and/or productivity.
- Better control over physiological responsiveness.
- Enhanced detection of cellular mediators.

Serum-free Culture

Overall, cells in serum-free culture are more sensitive to extremes of pH, temperature, osmolality, mechanical forces, and enzyme treatment.

- Antibiotics: It is best not to use antibiotics in serum-free media. If you do, we recommend that you use 5- to 10-fold less than you would in a serum- supplemented medium. This is because serum proteins tend to bind a certain amount of the antibiotic added; without these serum proteins the level of antibiotic may be toxic to certain cells.
- Higher Density: Cells must be in the mid-logarithmic phase of growth with viability >90% prior to adaptation. Sequential adaptation may be necessary. Seeding cultures at a higher density than normal at each passage during SFM adaptation may help the process. Because some percentage of cells may not survive in the new culture environment, having more cells present will increase the number of viable cells to further passage.
- Clumping: Cell clumping often occurs during adaptation to SFM. We recommend that you gently triturate the clumps to break them up when passaging cells.
- Morphology: It is not uncommon to see slight changes in cellular morphology during and after adaptation to SFM. As long as doubling times and viability remain good, slight changes in morphology should not be a reason for concern.

Basic Formula of Serum-free Medium

SFM includes basic culture medium and supplements. Basic culture mediums commonly use HamF12 and DMEM as 1:1 mixture.

The supplements include:

- Promote Adherence Substances: Many cells cultured in vitro have to attach the culture vessel for growth and SFM need to add some supplements for promoting attachment and extension, mainly extracellular matrix (fibronection, laminin, etc).
- Somatomedin and Hormones: Different growth factors need to add different cells. For example, EGF for the Keratinocytes, NGF for the Neurocyte, ECGF for the endothelial cell.
- Enzyme Inhibitor: Adherent cell need to trypsinize and passage. The SFM must have enzyme inhibitor to stop the activity of trypsin and protect the cell. The commonest enzyme inhibitor is soybean trypsin inhibitor.
- Binding Proteins and Translocator: The common binding protein and translocators are transferring and bovine serum albumin. Trace element. Selenium is most common trace element.

How to use

Sequential adaptation is preferred method for adapting cells to serum-free media (SFM).Because the change from 75% to 100% SFM may be too stressful for your cells, you may need to carry the cells for 2–3 passages in a 10% serum-supplemented medium: 90% SFM mixture. Most cell lines can be considered fully adapted after 3 passages in 100% SFM. Occasionally you may have trouble getting your cells past a certain step even before going 100% SFM. If this happens, go back and passage the cells 2–3 times in the previous ratio of serum-supplemented media to serum-free media.

Table: The utilization of some growth factor.

Growth factor	Target cell(common)	Target cell(specific)	Recommended concentration
EGF	epiblast, mesoblastema	Keratinocytes, fibroblast, chondrocyte, etc.	1-20ng/ml
bFGF	Mesoblastema, neuroectoderm	Endothelial cell, fibroblast, chondrocyte, myoblast, etc.	0.5-10ng/ml
FGF	Mesoblastema, neuroectoderm	Fibroblast, vascular cells.	1-100 ng/ml
ECGF	endothelial cell	Endothelial cell	1-3mg/ml
IGF-1	Most cells		1-10ng/ml
PDGF	mesenchymal cell	Fibroblast, myocyte, neurogliocyte	1-50ng/ml
NGF	sensory cell, sympathetoblast	Neurocyte, neurogliocyte	5-100ng/ml
TGF-α	Stimulate mesenchymal cell		0.1-3ng/ml
TGF-β	Inhibit epiblast		

Protein Free Medium (PFM) and Chemical Defined Medium (CDM)

Protein free medium (PFM) is a proprietary serum-free and protein-free growth medium that does not contain any hormones or growth factors. This medium is optimized for the cultivation of Chinese Hamster Ovary (CHO) cells in addition to many derivatives of this parent line. The absence of protein in the medium eliminates any transmission risk of blood borne diseases, resulting in a product that is far safer than the current culture media on the market. Chemical defined medium (CDM) is a media in which the chemical nature of all the ingredients and their amounts are known.

Other Medium

Balanced Salt Solution (BSS)

Balanced salt solutions can provide an environment that maintains the structural and physiological integrity of cells in vitro. Solutions most commonly include sodium, potassium, calcium, magnesium, and chloride. Balanced salt solutions are used for washing tissues and cells and are usually combined with other agents to treat the tissues and cells. They provide the cells with water and inorganic ions, while maintaining a physiological pH and osmotic pressure. The following table lists the formula of several common BSS.

Table: The formula of several common BSS.

	Ringer	PBS	Tyrode	Earle	Hank's	D-Hank's	Dulbecco
NaCl	9.00	8.00	8.00	6.80	8.00	8.00	8.00
KCl	0.42	0.20	0.20	0.40	0.40	0.40	0.20
$CaCl_2$	0.25	-	0.20	0.20	0.14	-	0.10
$MgCl_26H_2O$	-	-	0.10	-	-	-	0.10
$MgSO_47H_2O$	-	-	-	0.20	0.20	-	-
$Na_2HPO_4H_2O$	-	1.56	-	-	0.06	0.06	-
$NaH_2PO_42H_2O$	-	-	0.05	0.14	-	-	1.42
KH_2PO_4	-	0.20	-	-	0.06	0.06	0.20
$NaHCO_3$	-	-	1.00	2.20	0.35	0.35	-
Glucose	-	-	1.00	1.00	1.00	-	-
Phenol red	-	-	-	0.02	0.02	0.02	0.02

Medium used for Digestion

Primary cell cultures usually need to digest and dissociate the tissue to cell suspension. Continuous passage cultures need to digest the adherent cell from culture dishes. The common digestion solutions are trypsin, EDTA solution and collagenase solution.

Trypsin

In a tissue culture lab, trypsins are used to re-suspend cells adherent to the cell culture dish wall during the process of harvesting cells. Trypsin is used to cleave proteins bonding the cultured cells to the dish, so that the cells can be suspended in fresh solution and transferred to fresh dishes.

Trypsin is an endopeptidase produced by the gastro-intestines of mammals, and has an optimal operating pH of about 8 and an optimal operating temperature of about 37 °C. The biochemical assays performed on Trypsin 1:250 determine both trypsin specific-activity at the level of certain co-purified enzymes that influence cell removal and viability.

EDTA Solution

In Cell Culture applications, EDTA is used for its chelating properties which binds to calcium and other ions and thus prevents adjoining of cadherins (i.e. the integral membrane proteins involved in calcium-dependent cell-adhesion) between cells, preventing the clumping of cells growing in liquid suspension or even detaching adherent cells for passaging. The working concentration is 0.02%. Trypsin//EDTA is a combined method for detaching cells.

Collagenase Solution

Collagenase is especially valuable when tissues are too fibrous or too sensitive to allow the use of trypsin, which is ineffective on fibrous material and damaging to sensitive material. Dissociation is usually achieved either by perfusing whole organs or by incubating smaller pieces of tissue with enzyme solution. The working solution is 0.1-0.3mg/ml or 200000U/L. The optimal operating pH is about 6.5.

pH Adjusting Medium

$NaHCO_3$ Solution

Sodium bicarbonate is used as part of a buffering system commonly used to maintain physiological pH 7.2 - 7.4 in a culture environment. The concentration of the sodium bicarbonate in the medium must be matched with the level of CO_2 in the atmosphere above the medium. For media containing 1.5 to 2.2 g/L sodium bicarbonate, use 5% CO_2. For media containing 3.7 g/L sodium bicarbonate, use 10% CO_2.

HEPES

HEPES (4-(2-hydroxyethyl)-1-piperazineethanesulfonic acid) is a zwitterionic organic chemical buffering agent. The level of HEPES in cell culture media may vary from 10mM to 25mM. It is added to the media solely for extra buffering capacity when cell culture requires extended periods of manipulation outside of a CO_2 incubator.

Antibiotics

Researchers use antibiotics and antimycotics to prevent contamination. The common ones are Penicillin & Streptomycin (P/S). Penicillin Streptomycin mixtures contain 5,000 units of penicillin (base) and 5,000 µg of streptomycin (base)/ml utilizing penicillin G (sodium salt) and streptomycin sulfate in 0.85% saline. The following table summarizes the utilization of different antibiotics.

Table: Antibiotics Storage Conditions Working Concentration Stability in Culture (37 °C).

Antibiotic	Storage Conditions	Working Concentration	Stability in Culture (37 °C)
Amphotericin B	–20°C	2.5 µg/ml	~3 days
Ampicillin	4°C	100 µg/ml	~3 days
Chloramphenicol	–20°C	5 µg/ml	~5 days
Ciprofoxacin	–20°C	10 µg/ml	not known
Gentamicin	4°C	50 µg/ml	~5 days
Hygromycin B	4°C	500 µg/ml	not known
Kanamycin	–20°C	100 µg/ml	~5 days

Neomycin	−20°C	50 μg/ml	~5 days
Penicillin	−20°C	100 U/ml	~3 days
Puromycin	−20°C	20 μg/ml	not known
Streptomycin	−20°C	100 μg/ml	~3 days
Tetracycline	−20°C	10 μg/ml	~4 days

Glutamine

Glutamine is an unstable essential amino acid required in cell culture media formulations. Most commercially available media are formulated with free L-glutamine which is either included in the basal formula or added to liquid formulations at time of use. The concentration of L-glutamine used in classical media is usually 0.002mol/L.

Laboratory Requirements

Animal cell culture technique is one of the major techniques used in modern biology. Animal cell culture is an excellent model for studying the normal physiology and biochemistry of cells. Animal cell culture is also utilized for virus isolation, vaccine production as well as for study of toxicity, mutagenesis and carcinogenesis. This technique is also use for drug screening and manufacturing of biological compounds. Many therapeutic importance products can be manufacture with the help of animal cell culture technology. Whenever sterility chain break, it would lead to contamination in animal culture, that is not only wastage of money but dangerous also. Therefore, utmost care required is expected for animal cell culture. Although requirements of an animal cell culture laboratory depend mainly on the types of research or work in the mandate of laboratory. Manual handling should be avoided in animal cells culture laboratory because it may be led to contamination. General microbiology principle should be follow and work should do into designated bio-safety level. Ethical issue should be considering while working with animal cell culture laboratory. All animal cell culture laboratories require same basic equipment that is essential for smooth and contamination free culture. Specifically, we require the instruments for aseptic condition, incubation and storage of animal cell culture.

Types of the Requirements for Animal Cell Culture Laboratory:

- Physical Requirements,
- Substrates for Animal cell culture Growth,
- Biological components.

Table: Physical Requirements for Animal Cell Culture Laboratory.

S. No.	Requirement	Function
1	Air-conditioning of a room/space	Help in Incubation
2	Hot room with temperature recorder	Dryness of material
3	Microscope room	Microscopic works
4	Dark room	Sterilization of glassware & culture media etc.
5	Sterilization room	
6	Media preparation room	For media preparation & sterilization

Storage areas		
1	Freeze (4°C)	Store the reagents & chemical
2	Freeze (20°C)	Store the reagents & chemical
3	Deepfreeze (76-86°C)	Serum, sample, DNA, RNA
4	Liquid Nitrogen	Cell-line
5	Glassware	Reagent bottle
6	Plastics	T-25 flask, culture plate
7	Small items-drawers,	Tips, tube & cryo-vial
8	Filter assembly	Filter assembly & filter
9	Specialized equipment cupboard	Sterile Glassware & Plastics
10	Chemicals-sidled containers	Reagents & chemical

Physical Requirements

- Aseptic Condition: Aseptic condition is required to produce contamination free cells. Either separate sterile room should be available or a designated area within a larger laboratory for sterile handling, incubation and storage of cell cultures, reagents and media. All aseptic work should be done under laminar airflow or biosafety cabinets. There should be minimum movement in tissue culture room while the work on progress. The person should be expert in microbiology for safely handling the material in animal cells culture laboratory. The person should wear mask, glows, lab-coat/ gown while working in the animal cells culture laboratory. Animal cell culture laboratory waste material should be discarded after proper autoclave and following the biosafety principles.

- Incubator: For the specific growth of cell culture appropriate environment like temperature, humidity, CO_2 are required. These conditions can be hassle-free provided with the help of CO_2 incubator. CO_2 incubator also helpful in maintain of the pH and improve the growth of the cells. CO_2 incubator should be water jacketed or there should be stainless-steel trays with neat and clean double distil water inside the CO_2 incubator. Infected or sterile culture should be incubated separately to avoid the cross contamination. CO_2 incubator should be able to maintain of accurate temperature and CO_2 level.

- Storage: Animal cell culture laboratory should have neat and clean storage areas for media, reagents, drugs and antibiotics at appropriate temperature condition. Storage is required for disposable pipettes, culture vessels, gloves, glassware such as media bottles and glass pipettes. All the things should store as per standard tissue culture protocol and manufacturer instructions. All disposable pipettes, culture vessels, gloves, glassware such as media bottles and glass pipettes should be purchase from the authentic source only. Expiry dates reagents should not use in the animal cell culture laboratory.

- Refrigerators: Most cell culture reagents can be stored at -5°-20°C. A domestic refrigerator and -20°C required for storing the reagents and media at -5°-20°for small laboratory while a cold room for large laboratory. All the things should store as per standard tissue culture protocol and manufacturer instructions at the specified temperature. The things should be proper label. There should be proper and continuous power supply to the refrigerator. Make sure that the refrigerator or the cold room is cleaned regularly to avoid contamination.

Cell Culture Growing Material

There are two categories of cells: suspension & anchorage dependent. The suspension type of cells is floating into the media therefore no specific surface required. For the anchorage, dependent specific surface required for their attachment and growth. These surfaces may be e.g. plastic, glass, palladium, metallic surfaces, etc. Disposal and transparent plastic is the best option in the present scenario.

- Disposable Plastic as a Substrate: Disposable plastics are cheaper substrate as they are commonly made up of polystyrene. After using they should be thrown at proper place by following disposal rule. They are sterile with gamma the help of radiation or electric arc which is also develops charges on the surface of substrate. These are transparent; therefore, proliferation and cell growth can be easily measured. It should be noted that only monolayer of cell must be grown. Cell culture can be scale up with the help of plastic beads of polystyrene, sephadex.
- Glass as a Substrate: Glass is an important substrate used in laboratory in several forms such as test tubes, slides, coverslips, pipettes, flasks, rods, bottles, Petti dishes, several apparatuses etc. These are sterilized by using chemicals, radiations, dry beat (in oven) and moist heat. Glassware is more economical as compare with disposable plastics, if repeatedly used after sterilization with the help of autoclave and oven.
- Palladium as a Substrate: For the first-time palladium deposited on agarose was used as a substrate for growth of fibroblast and glia.

Biological Components

Cell-line can be obtained from the National Centre for Cell Science, Pune or any other research centre of world level reputation. After the authenticity of cell, proper medium, specified condition should be provided in sterile condition for their growth.

Cryopreservation

Cell lines in continuous culture are likely to suffer from genetic instability as their passage number increases. Low passage level cells are more effective as compare with high passage level cell. Therefore, it is advisable to prepare working stocks of the cells and preserve them at -80°C or liquid nitrogen with DMSO. For short term cell can be cryopreserved at -80°C while for long term cell can be cryopreserved at liquid nitrogen only. Repeated freezing and toughing of the animal cell culture should be avoided.

Types of Animal Cell Culture

Primary Cell Culture

Primary cell culture refers to growing and maintaining of the selected cell type which has been excised from the normal parental tissue. Primary cell culture provides more biologically relevant data than that generated using cell lines. Concerns over the use of cell lines have resulted in a growing need for primary cells in a variety of applications from basic research to drug discovery. Often, primary cells are combined with newer technologies such as 3D cell culture given a recent surge within the research community to use better reagents to improve research.

Primary culture broadly involves the culturing techniques carried following the isolation of the cells, but before the first subculture. Primary cultures are usually prepared from large tissue masses. Thus, these cultures may contain a variety of differentiated cells e.g. fibroblasts, lymphocytes, macrophages, epithelial cells.

With the experiences of the personnel working in tissue culture laboratories, the following criteria/ characteristics are considered for efficient development of primary cultures:

- Embryonic tissues rather than adult tissues are preferred for primary cultures. This is due to the fact that the embryonic cells can be disaggregated easily and yield more viable cells, besides rapidly proliferating in vitro.
- The quantity of cells used in the primary culture should be higher since their survival rate is substantially lower (when compared to subcultures).
- The tissues should be processed with minimum damage to cells for use in primary culture. Further, the dead cells should be removed.
- Selection of an appropriate medium (preferably a nutrient rich one) is advisable. For the addition of serum, fetal bovine source is preferred rather than calf or horse serum.
- It is necessary to remove the enzymes used for disaggregation of cells by centrifugation.

Techniques for Primary Culture

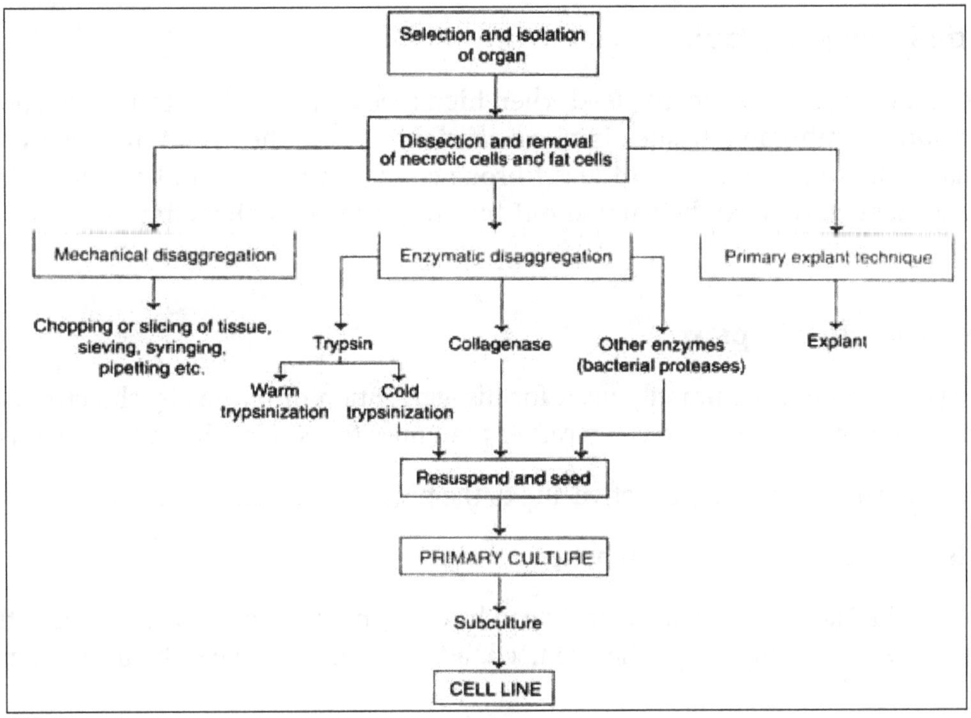

Figure: Different techniques used for primary culture.

Among the various techniques devised for the primary culture of isolated tissues, three techniques are most commonly used:

- Mechanical disaggregation,

- Enzymatic disaggregation,
- Primary explant technique.

An outline of these techniques is depicted in figure, and the procedures are briefly described:

Mechanical Disaggregation

For the disaggregation of soft tissues (e.g. spleen, brain, embryonic liver, soft tumors), mechanical technique is usually employed. This technique basically involves careful chopping or slicing of tissue into pieces and collection of spill out cells.

The cells can be collected by two ways:

- Pressing the tissue pieces through a series of sieves with a gradual reduction in the mesh size.
- Forcing the tissue fragments through a syringe and needle.

Although mechanical disaggregation involves the risk of cell damage, the procedure is less expensive, quick and simple. This technique is particularly useful when the availability of the tissue is in plenty, and the efficiency of the yield is not very crucial. It must however, be noted that the viability of cells obtained from mechanical techniques is much lower than the enzymatic technique.

Enzymatic Disaggregation

Enzymatic disaggregation is mostly used when high recovery of cells is required from a tissue. Disaggregation of embryonic tissues is more efficient with higher yield of cells by use of enzymes. This is due to the presence of less fibrous connective tissue and extracellular matrix. Enzymatic disaggregation can be carried out by using trypsin, collagenase or some other enzymes.

Disaggregation by Trypsin

The term trypsinization is commonly used for disaggregation of tissues by the enzyme, trypsin. Many workers prefer to use crude trypsin rather than pure trypsin for the following reasons:

- The crude trypsin is more effective due to the presence of other proteases.
- Cells can tolerate crude trypsin better.
- The residual activity of crude trypsin can be easily neutralized by the serum of the culture media (when serum-free media are used, a trypsin inhibitor can be used for neutralization).

Disaggregation of cells can also be carried out by using pure trypsin which is less toxic and more specific in its action. The desired tissue is chopped to 2-3 mm pieces and then subjected to disaggregation by trypsin. There are two techniques of trypsinization-warm trypsinization and cold trypsinization.

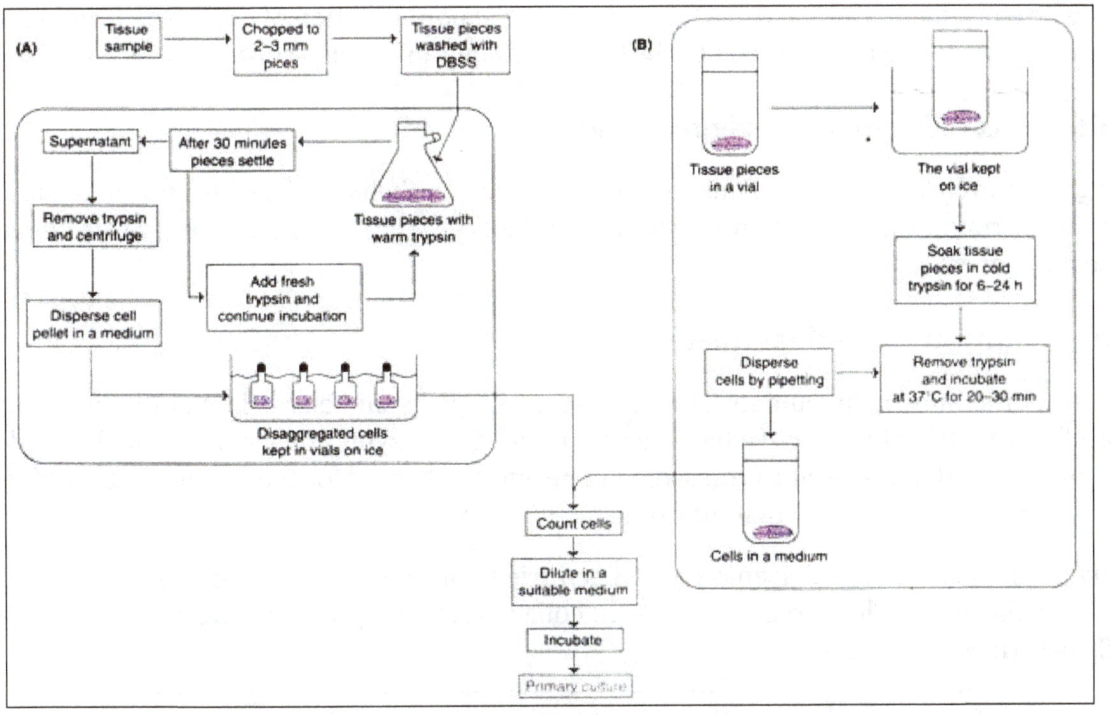

Figure: Preparation of primary culture by trypsin disaggregation (A) Warm trypsinization (B) Cold trypsinization (DBSS-Dissection basal salt solution).

Warm Trypsinization

This method is widely used for disaggregation of cells. The chopped tissue is washed with dissection basal salt solution (DBSS), and then transferred to a flask containing warm trypsin (37° C). The contents are stirred, and at an interval of every thirty minutes, the supernatant containing the dissociated cells can be collected. After removal of trypsin, the cells are dispersed in a suitable medium and preserved (by keeping the vial on ice).

The process of addition of fresh trypsin (to the tissue pieces), incubation and collection of dissociated cells (at 30 minutes intervals) is carried out for about 4 hours. The disaggregated cells are pooled, counted, appropriately diluted and then incubated.

Cold Trypsinization

This technique is more appropriately referred to as trypsinization with cold pre-exposure. The risk of damage to the cells by prolonged exposure to trypsin at 37°C (in warm trypsinization) can be minimized in this technique.

After chopping and washing, the tissue pieces are kept in a vial (on ice) and soaked with cold trypsin for about 6-24 hours. The trypsin is removed and discarded. However, the tissue pieces contain residual trypsin. These tissue pieces in a medium are incubated at 37°C for 20-30 minutes. The cells get dispersed by repeated pi-pettings. The dissociated cells can be counted, appropriately diluted and then used.

The cold trypsinization method usually results in a higher yield of viable cells with an improved survival of cells after 24 hours of incubation. This method does not involve stirring or centrifugation, and can be conveniently adopted in a laboratory. The major limitation of cold trypsinization is that it is not suitable for disaggregation of cells from large quantities of tissues.

Limitations of Trypsin Disaggregation

Disaggregation by trypsin may damage some cells (e.g. epithelial cells) or it may be almost ineffective for certain tissues (e.g. fibrous connective tissue). Hence other enzymes are also in use for dissociation of cells.

Disaggregation by Collagenase

Collagen is the most abundant structural protein in higher animals. It is mainly present in the extracellular matrix of connective tissue and muscle. The enzyme collagenase (usually crude one contaminated with non-specific proteases) can be effectively used for the disaggregation of several tissues (normal or malignant) that may be sensitive to trypsin.

Highly purified grades of collagenase have been tried, but they are less effective when compared to crude collagenase. The important stages in collagenase disaggregation, depicted in Figure, are briefly described hereunder.

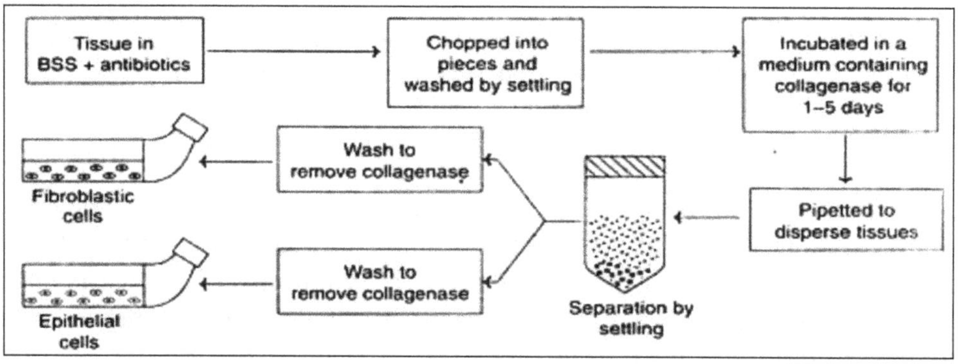

Figure: Important stages in collagenase disaggregation of tissue for primary culture (BSS-Basal Salt solution).

The desired tissue suspended in basal salt solution, containing antibiotics is chopped into pieces. These pieces are washed by settling, and then suspended in a complete medium containing collagenase. After incubating for 1-5 days, the tissue pieces are dispersed by pipetting. The clusters of cells are separated by settling. The epithelial cells and fibroblastic cells can be separated. Collagenase disaggregation has been successfully used for human brain, lung and several other epithelial tissues, besides various human tumors, and other animal tissues. Addition of another enzyme hyaluronidase (acts on carbohydrate residues on cell surfaces) promotes disaggregation.

Collagenase in combination with hyaluronidase is found to be very effective for dissociating rat or rabbit liver. This can be done by per-fusing the whole organ in situ. Some workers use collagenase in conjunction with trypsin, a formulation developed in chick serum, for disaggregation of certain tissues.

Use of other Enzymes in Disaggregation

Trypsin and collagenase are the most widely used enzymes for disaggregation. Certain bacterial proteases (e.g. pronase, dispase) have been used with limited success. Besides hyaluronidase, neuraminidase is also used in conjunction with collagenase for effective degradation of cell surface carbohydrates.

Primary Explant Technique

The primary explant technique was, in fact the original method, developed by Harrison in 1907. This technique has undergone several modifications, and is still in use. The simplified procedure adopted for primary explant culture is depicted in figure, and briefly described below:

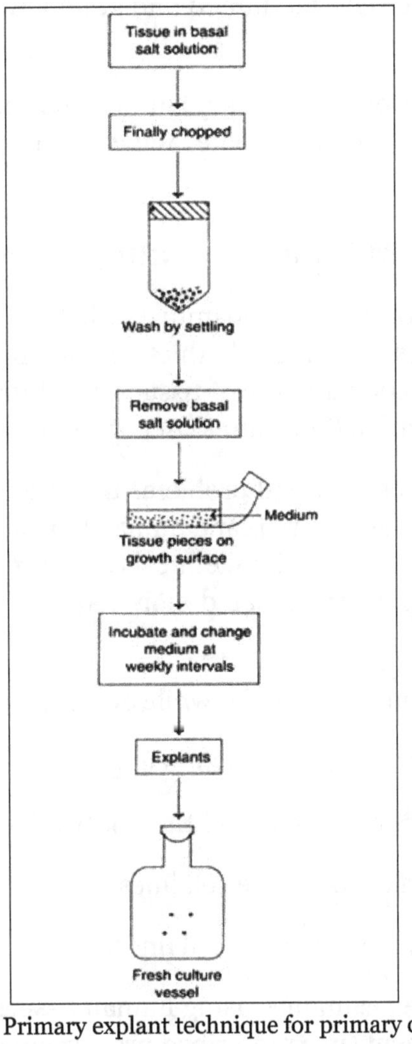

Figure: Primary explant technique for primary culture.

The tissue in basal salt solution is finely chopped, and washed by settlings. The basal salt solution is then removed. The tissue pieces are spread evenly over the growth surface. After addition of appropriate medium, incubation is carried out for 3-5 days. Then the medium is changed at weekly intervals until a substantial outgrowth of cells is observed. Now, the explants are removed and transferred to a fresh culture vessel.

The primary explant technique is particularly useful for disaggregation of small quantities of tissues (e.g. skin biopsies). The other two techniques mechanical or enzymatic disaggregations however, are not suitable for small amounts of tissues, as there is a risk of losing the cells.

The limitation of explant technique is the poor adhesiveness of certain tissues to the growth surface, and the selection of cells in the outgrowth. It is however, observed that the primary explant technique can be used for a majority of embryonic cells e.g. fibroblasts, myoblasts, epithelial cells, glial cells.

Separation of Viable and Non-viable Cells

It is a common practice to remove the nonviable cells while the primary culture is prepared from the disaggregated cells. This is usually done when the first change of the medium is carried out. The very few left over non-viable cells get diluted and gradually disappear as the proliferation of viable cells commences.

Sometimes, the non-viable cells from the primary cultures may be removed by centrifugation. The cells are mixed with ficoll and sodium metrizoate, and centrifuged. The dead cells form a pellet at the bottom of the tube.

Medical Ethics and Safety Measures in Culture Techniques

Since the culture techniques involve the use of animal or human tissues, it is absolutely necessary to follow several safety measures and medical ethics. In fact, in some countries there are established legislation/norms for selection and use of tissues in cultures. For example, in United Kingdom, Animal Experiments (Scientific Procedures) Act of 1986 is followed.

The handling of human tissues poses several problems that are not usually encountered with animal tissues. While dealing with fetal materials and human biopsies, the consent of the patient and/ his or her relatives, besides the consent of local ethical committee is required. Further, taking any tissue (even in minute quantities) from human donors requires the full consent of the donor in a prescribed format.

The following issues need to be fully considered while dealing with human tissues:

- The consent of the patient and relatives for using tissues for research purposes.
- Ownership of the cell lines developed and their derivatives.
- Consent for genetic modification of the cell lines.
- Patent rights for any commercial use of cell lines.

In the general practice of culture techniques using human tissues, the donor and/or relatives are asked to sign a disclaimer statement (in a prescribed pro-forma) before the tissue is taken. By this approach, the legal complications are minimized.

Safety Measures

Handling of human tissues is associated with a heavy risk of exposure for various infections. Therefore, it is absolutely necessary that the human materials are handled in a biohazard cabinet. The tissues should be screened for various infections such as hepatitis, tuberculosis, HIV, before their use. Further, the media and apparatus, after their use must be autoclaved or disinfected, so that the spread of infections is drastically reduced.

Secondary Cell Culture

When primary culture is passaged or sub-cultured, it becomes secondary culture or cell line. Passaging or Subculture means removal of medium and transfer of cells from previous culture vessel to another fresh culture vessel with fresh growth medium. It is regularly required for the supply of fresh nutrients and increasing space for continuously dividing cells. Sub culturing of cells from one culture vessel to another involves disassociating the adhered cells by using enzyme trypsin and adding fresh growth media. It depends on the characteristics and growth rates of particular cell types to decide the subculture or splitting ratio.

Subculture of Cell

Subculture (or passage) refers to the transfer of cells from one culture vessel to another culture vessel. Subculture usually (not always) involves the subdivision of proliferating cells that enables the propagation of a cell line. The term passage number is used to indicate the number of times a culture has been sub-cultured.

The standard growth curve of cells in a culture is depicted in figure. The initial lag phase is followed by exponential or log phase and a plateau phase. During the active growth period in the log phase, the medium must be changed frequently, or else growth ceases. As the cell concentration exceeds the capacity of the medium, the culture has to be divided to subculture.

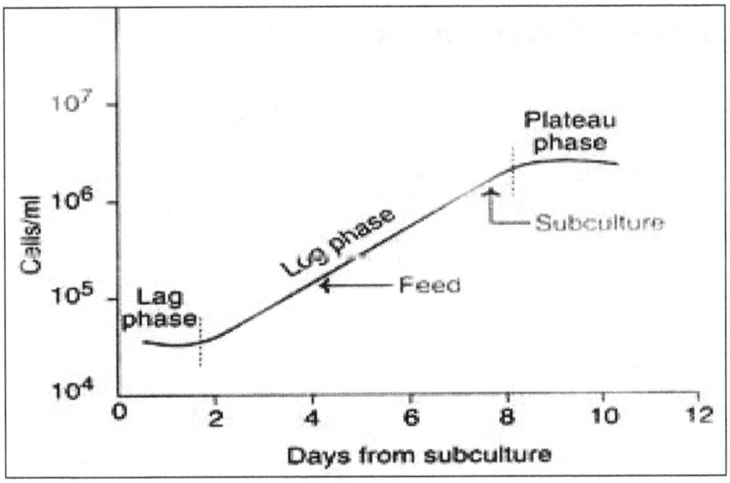

Figure: Depiction of a standard growth of cells in a culture.

There are two types of subcultures-monolayer subculture and suspension subculture:

Monolayer Cultures

When the bottom of the culture vessel is covered with a continuous layer of cells, usually one cell in thickness, they are referred to as monolayer cultures. The attachment of cells among themselves and to the substrate (i.e. culture vessel) is mediated through surface glycoproteins (cell adhesion molecules) and calcium ions (Ca^{2+}). For sub-culturing of monolayer cultures, it is usually necessary to remove the medium and dissociate the cells in the monolayer by degrading the cell adhesion molecules, besides removing Ca^{2+}.

Methods of Cell Dissociation

There are physical and enzymatic methods for dissociation of monolayer cultures.

Table: Commonly used cell dissociation methods for monolayer cultures

Method	Applicable to
Physical	
Mechanical shaking	Loosely adherent cells
Scraping	Cells sensitive to proteases
Enzymatic	
Trypsin cell lines	Most of the continuous
Trypsin + collagenase cultures	Dense and multilayer
Dispase	Removal of epithelium in sheets
Pronase	Single-cell suspensions

Mechanical shaking and cell scraping are employed for cultures which are loosely adhered, and the use of proteases has to be avoided. Among the enzymes, trypsin is the most frequently used. For certain cell monolayers, which cannot be dissociated by trypsin, other enzymes such as pronase, dispase and collagenase are used. Prior to cell dissociation by enzymes, the monolayers are usually subjected to pretreatment of EDTA for the removal of Ca^{2+}.

Criteria for Subculture of Monolayers

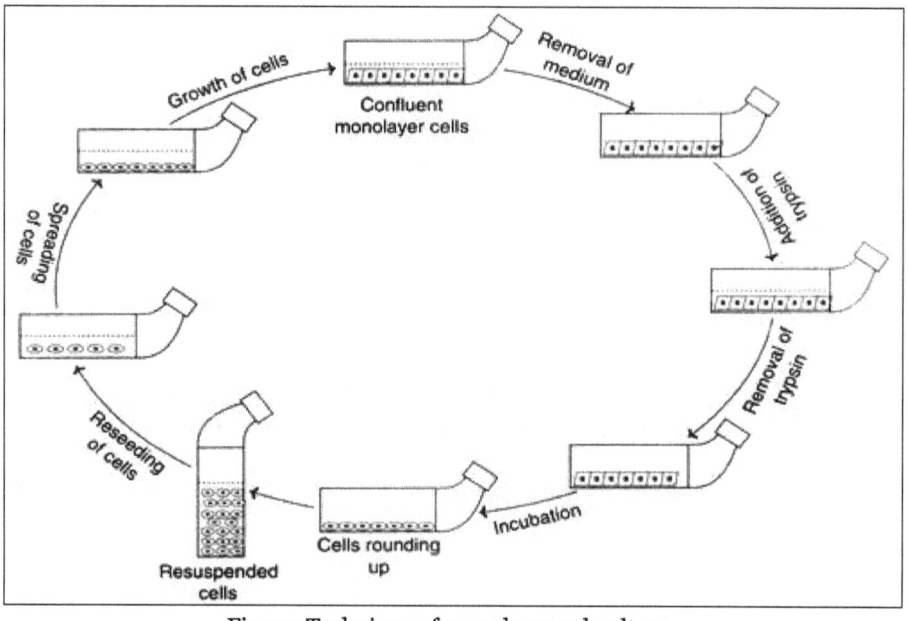

Figure: Technique of monolayer subculture.

The sub-culturing is ideally carried out between the middle of the log phase and the time before they enter plateau phase. Subculture of cells should not be done when they are in lag phase. The other important criteria for subculture of monolayers are briefly described.

- Culture Density: It is advisable to subculture the normal or transformed monolayer cultures, as soon as they reach confluence. Confluence denotes the culture stage wherein all the available growth area is utilized and the cells make close contact with each other.

- Medium Exhaustion: A drop in pH is usually accompanied by an increase in culture cell density. Thus, when the pH falls, the medium must be changed, followed by subculture.
- Scheduled Timings of Subculture: It is now possible to have specified schedule timings for subculture of each cell line. For a majority of cell cultures, the medium change is usually done after 3-4 days, and sub-culturing after 7 days.
- Purpose of Subculture: The purpose for which the cells are required is another important criterion for consideration of sub-culturing. Generally, if the cells are to be used for any specialized purpose, they have to be sub-cultured more frequently.

Techniques of Monolayer Subculture

The subculture of monolayer cells basically consists of the following steps:

- Removal of the medium.
- Brief exposure of the cells to trypsin.
- Removal of trypsin and dispersion in a medium.
- Incubation of cells to round up.
- Re-suspension of the cells in a medium for counting and reseeding.
- Cells reseeded and grown to monolayers.

Cell Concentration at Subculture

Most of the continuous cell lines are sub-cultured at a seeding concentration between 1×10^4 and 5×10^4 cells/ml. However for a new culture, subculture has to be started at a high concentration and gradually reduced.

Suspension Cultures

Majority of continuous cell lines grow as monolayers. Some of the cells which are non- adhesive e. g. cells of leukemia or certain cells which can be mechanically kept in suspension can be propagated in suspension. The transformed cells are sub-cultured by this method. Subculture by suspension is comparable to culturing of bacteria or yeast.

Advantages of Cell Propagation by Suspension

- The process of propagation is faster.
- The lag period is usually shorter.
- Results in homogeneous suspension of cells.
- Treatment with trypsin is not required.

- Scale-up is convenient.
- No need for frequent replacement of the medium.
- Maintenance is easy.
- Bulk production of cells can be conveniently achieved.

Criteria for Suspension Subculture

The criteria adopted for suspension subculture are the same as that already described for monolayer subcultures. The following aspects have to be considered:

- Culture density.
- pH change representing medium exhaustion.
- Schedule timings of subculture.
- Purpose of subculture.

Technique of Suspension Culture

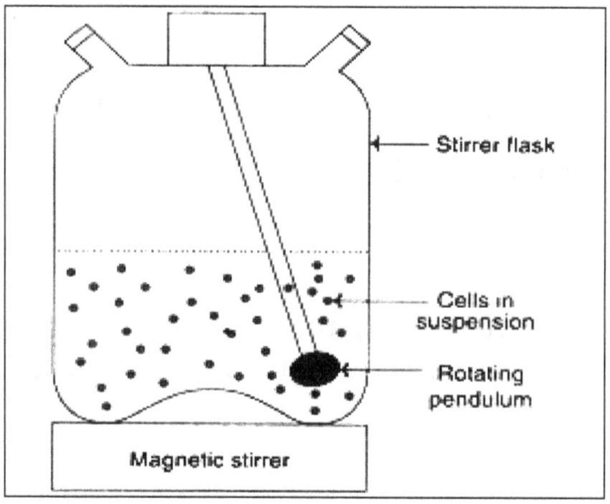

Figure: Suspension culture in a stirrer flask.

The cells can be suspended in a culture flask (a stirrer flask) containing the desired medium. The medium is continuously stirred with a magnetic pendulum rotating at the base of the flask. The cells have to periodically examine for contamination or signs of deterioration.

Cell Lines

A cell line or cell strain arises from primary or secondary cultures at the time of sub culturing of cells. The term cell line depicts that cultured cells consist of cell that were originally present in primary culture. On the basis of the life span of culture, the cell lines are categorized into two types - finite or continuous depending upon whether it has limited culture life span or it is immortal in culture.

- Finite Cell Lines: Cell lines which grows for a limited life span and have limited number of cell divisions (around 20 - 80 population doublings or subcultures) are known as finite cell lines. The factors which manage the division of these cells in vitro are related to the type and differentiation level of the starting cell type. The cell types will proliferate for an absolute period after which culture will ultimately stops dividing. These cells can be used for the generation large number of cells without immortalization procedures as well as generation of stably transfected clones.

- Continuous Cell Lines: Those cell line which gets transformed or immortalized either automatically or manually during subcultures under in vitro culture conditions changes into continuous cell lines. These types of cell lines gains potential for sub culturing indefinitely, becomes anchorage independent, lack contact inhibition, and accumulate aneuploidy or heteroploidy. These cell lines are known as Infinite or immortal cell lines. These cells can be grown in culture for a very long duration with fast growth rates and 12 hrs-24hrs of doubling times.

Table: Comparison of properties of finite and continuous cell lines.

Property	Finite cell line	Continuous cell line
Growth rate	Slow	Fast
Mode of growth	Monolayer	Suspension of monolayer
Yield	Low	High
Transformation	Normal	Immortal, tumorigenic
Ploidy	Euploid (multiple of haploid chromosomes)	Aneuploid (not an exact multiple of haploid chromosomes)
Anchorage dependence	Yes	No
Contact inhibition	Yes	No
Cloning efficiency	Low	High
Serum requirement	High	Low
Markers	Tissue specific	Chromosomal, antigenic or enzymatic

The most commonly used terms while dealing with cell lines are explained below:

- Split ratio: The divisor of the dilution ratio of a cell culture at subculture. For instance, when each subculture divided the culture to half, the split ratio is 1:2.

- Passage number: It is the number of times that the culture has been sub-cultured.

- Generation number: It refers to the number of doublings that a cell population has undergone. It must be noted that the passage number and generation number are not the same, and they are totally different.

Nomenclature of Cell Lines

It is a common practice to give codes or designations to cell lines for their identification. For instance, the code NHB 2-1 represents the cell line from normal human brain, followed by cell strain (or cell line number) 2 and clone number 1. The usual practice in a culture laboratory is to maintain a log book or computer database file for each of the cell lines.

While naming the cell lines, it is absolutely necessary to ensure that each cell line designation is unique so that there occurs no confusion. Further, at the time of publication, the-cell line should be prefixed with a code designating the laboratory from which it was obtained e.g. NCI for National Cancer Institute, Wl for Wistar Institute.

Commonly used Cell Lines

There are thousands of cell lines developed from different laboratories world over. A selected list of some commonly used cell lines along with their origin, morphology and other characters are given in table.

Table: A selected list of commonly used cell lines.

Cell line	Species of origin	Tissue of origin	Morphology	Ploidy	Characteristics
IMR-90	Human	Lung	Fibroblast	Diploid	Susceptible to human viral infections.
3T3-A31	Mouse	Connective tissue	Fibroblast	Aneuploid	Contact inhibited, readily transformed
BHK21-C13	Hamster (Syrian)	Kidney	Fibroblast	Aneuploid	Readily transformable
CHO-k1	Chinese hamster	Ovary	Fibroblast	Diploid	Simple karyotype
NRK49F	Rat	Kidney	Fibroblast	Aneuploid	Induction of suspension growth by TGF- α, β.
BRL 3A	Rat	Liver	Epithelial	Diploid	Produces IGF-2
Vero	Monkey	Kidney	Fibroblast	Aneuploid	Viral substrate and assay
HeLa-S_3	Human	Cervical carcinoma	Epithelial	Aneuploid	Rapid growth, high plating efficiency
Sk/HEP-I	Human	Hepatoma	Endothelial	Aneuploid	Factor VIII
Caco-2	Human	Colo-rectal carcinoma	Epithelial	Aneuploid with polarised	Forms tight monolayer support
MCF-7	Human	Breast tumor (effusion)	Epithelial	Aneuploid	Estrogen receptor positive
Friend	Mouse	Spleen	Suspension	Aneuploid	Hemaglobin, growth hormone.

Selection of Cell Lines

Several factors need to be considered while selecting a cell line. Some of them are briefly described:

- Species: In general, non-human cell lines have less risk of biohazards, hence preferred. However, species differences need to be taken into account while extrapolating the data to humans.
- Finite or continuous cell lines: Cultures with continuous cell lines are preferred as they grow faster, easy to clone and maintain, and produce higher yield. But it is doubtful whether the continuous cell lines express the right and appropriate functions of the cells. Therefore, some workers suggest the usc of finite cell lines, although it is difficult.
- Normal or transformed cells: The transformed cells are preferred as they are immortalized and grow rapidly.

- Availability: The ready availability of cell lines is also important. Sometimes, it may be necessary to develop a particular cell line in a laboratory.
- Growth characteristics: The following growth parameters need to be considered:
    - Population doubling time.
    - Ability to grow in suspension.
    - Saturation density (yield per flask).
    - Cloning efficiency.
- Stability: The stability of cell line with particular reference to cloning, generation of adequate stock and storage are important.
- Phenotypic expression: It is important that the cell lines possess cells with the right phenotypic expression.

Maintenance of Cell Cultures

For the routine and good maintenance of cell lines in culture (primary culture or subculture) the examination of cell morphology and the periodic change of medium are very important.

Cell Morphology

The cells in the culture must be examined regularly to check the health status of the cells, the absence of contamination, and any other serious complications (toxins in medium, inadequate nutrients etc).

Replacement of Medium

Periodic change of the medium is required for the maintenance of cell lines in culture, whether the cells are proliferating or non-proliferating. For the proliferating cells, the medium need to be changed more frequently when compared to non-proliferating cells. The time interval between medium changes depends on the rate of cell growth and metabolism.

For instance, for rapidly growing transformed cells (e.g. HeLa), the medium needs to be changed twice a week, while for slowly growing non-transformed cells (e.g. IMR-90) the medium may be changed once a week. Further, for rapidly proliferating cells, the sub-culturing has to be done more frequently than for the slowly growing cells.

The following factors need to be considered for the replacement of the medium:

- Cell concentration: The cultures with high cell concentration utilize the nutrients in the medium faster than those with low concentration; hence the medium is required to be changed more frequently for the former.
- A decrease in pH: A fall in the pH of the medium is an indication for change of medium. Most of the cells can grow optimally at pH 7.0, and they almost stop growing when

the pH falls to 6.5. A further drop in pH (between 6.5 and 6.0), the cells may lose their viability.

- The rate of fall in pH is generally estimated for each cell line with a chosen medium. If the fall is less than 0.1 pH units per day, there is no harm even if the medium is not immediately changed. But when the fall is 0.4 pH units per day, medium should be changed immediately.

Cell Type: Embryonic cells, transformed cells and continuous cell lines grow rapidly and require more frequent sub-culturing and change of medium. This is in contrast to normal cells, which grow slowly.

- Morphological changes: Frequent examination of cell morphology is very important in culture techniques. Any deterioration in cell morphology may lead to an irreversible damage to cells. Change of the medium has to be done to completely avoid the risk of cell damage.

Organ Culture

Organ culture means the explanation and growth in vitro of organs or part of organs in which the various tissue components, such as parenchyma and stroma, and their anatomical relationship and function, are preserved in culture, so that the explanted tissue closely resembles its parent tissue in vivo.

The outgrowth of isolated cells from the periphery of the explants is discouraged and minimised by suitable culture conditions, and the 'new' growth is composed of differentiated structures. Thus, in glands, new glandular structures are formed, or, in lung tissue, new small bronchi develop at the periphery of the explant. They consist of alveoli lined with secretory, cuboidal, or columnar glandular or bronchial epithelium.

In tissues lined with squamous epithelium, such as skin or oesophagus, or in bladder lined with transitional epithelium, the epithelium follows a similar pattern of differentiation as in the organs in vivo. Hormone-dependent tissues remain hormone sensitive and responsive, and endocrine organs continue to secrete specific hormones. Finally, in fetal tissues, morphogenesis in vitro closely resembles that seen in vivo.

Requirements for Organ Culture

Sterile Work Area

Working with culturing technique requires a separate clean room. Latter should be equipped with a laminar air flow cabinet which supplies filtered air through a HEPA (High Efficiency Particle Air Filter) filter. This facility should be used specifically for desired cell or tissue culture work. There should not be any interference from other lab practices like micro-organisms culturing.

Incubation Facilities

For better growth of cells in culture, laboratory should have CO_2 incubator maintained at 30-37°C.

Incubator should be designated only to work with cell types from mammals and insects. The incubation temperature always depends upon cell type being cultured. Insect cells grow best at around 30°C while mammalian cells require a temperature of 37°C .

Refrigerators and Freezer

A designated refrigeration facility is required for storage of liquid media at 4°C and other essential components like enzymes, amino acids, serum and antibiotics at -20°C. It is always a good thing to have pre aliquot stocks of serum, nutrients and antibiotics stored at required temperature. For preservation of cells/cell line it is mandatory to have liquid nitrogen facility or a -70°C freezer.

Sterilization

The sterilization is fundamental requirement for culturing of cell, tissue & organ. Primary purpose of sterilization is to kill the microorganisms. It can be achieved by following means:

- Dry heat: It sterilizes materials by denaturing proteins in microorganism that cause free radical damage which leads to death of microbial cells. Sterilization is achieved at minimum temperature of 160°C for two hours. Various instruments like hot air oven, microwave, Bunsen burner and incineration are used for dry heat sterilization.
- Moist heat: It is also one of efficient method for achieving sterilization. It is preferable method for sterilization of glass wares and certain types of plastic wares. It is achieved by using instrument like autoclave maintained at 121°C for 15-20 minutes.
- Filtration: Culture media, enzymes, hormones, cofactors and bicarbonate buffers cannot be sterilized either by dry heat or moist heat. Their sterilization is achieved by filtration process in which we passes out media or other biological fluids through a 0.22 µm pore size sterile and disposable membrane filter. These filters are consisting of materials like nylon, cellulose acetate, cellulose nitrate, polycarbonate, and polyethersulfone (PES).

Microscopy

Microscope especially inverted is essential requirement for examination of cultured cells, tissue and organ in flasks and dishes. The microscope detects morphological changes in cultures as they provide first indication of deterioration of a culture.

Culture Ware

In last few decades, glass wares were used as first choice material for cell culturing. However, with passage of time and development of more suitable & reliable alternatives such as plastic wares, they have been almost discontinued now. A variety of plastic wares being used nowadays. The most commonly used plastics wares are consisting of materials like polystyrene, polyvinyl chloride (PVC), polycarbonate and metinex. Cells are usually maintained in multiwall plates, stirrer bottles and petri dishes or flasks (25 cm^2 or 75 cm^2) with or without perforated cap. The actual choice of culture ware is affected by following factors:

- Culture type i.e. monolayer or suspension.

- Number of cells required.
- The objective.
- Cost of expenditure.

Other Requirements

- Temperature: It has been observed that temperature plays a critical role in growth of mammalian cells in culture. The recommended optimum temperature for incubation is 37°C, similar one to temperature of human body.
- Culture media: The culture media are essential for growth of cells as it provides:
  - The optimum conditions for pH and osmotic pressure.
  - Chemical constituents which are not synthesized by cultured cells.

Generally media contains organic salts, amino acids, fatty acids, sugars, ions, trace elements, vitamins and cofactors. Glucose can be added as energy source. Phenol red acts as pH indicator in medium. Examples of commonly used media includes minimal essential medium (MEM), RPMI 1640 medium, CMRL 1066 and F12. Culture media for animal cells and tissues can be divided into:

- Natural media: It represents natural sources that contains nutrient in sufficient amount for the growth and proliferation of animal cells and tissues.
- Synthetic media: It is prepared artificially by adding several organic and inorganic nutrients, vitamins, salts, serum proteins, carbohydrates and co-factors. A variety of synthetic media are available but they are broadly classified into:
  - Serum containing media (media containing serum).
  - Serum- free media (media without serum).

Characteristics of Organ Culture

The organ culture technique has become popular in last decades due to following characteristics:

Gas and Nutrient Exchange

During culturing of cells in the form of solid mass of tissue, the size of tissue is limited by gaseous diffusion rate and exchange of metabolite and nutrients. Cell survival in suspension culture or as monolayer is affected by dimension (250 μm) of aggregated cells. Latter also can limit diffusion rate or some time even cause necrosis of growing cells. To overcome this problem, organ cultures are preferred as they are placed at gas-liquid interphase, to facilitate gaseous exchange and access to nutrients. Major advantage of culture at gas-liquid interface is that the explant retains a spherical geometry.

Structural Integrity

In comparison to cell culture, organ culture is preferred in vitro technique because of its structural integrity. In cell culture, cells tend to dissociate during mechanical and enzymatic treatment while organ culture maintains the cellular associations.

Growth and Differentiation

A relationship exists between growth and differentiation as differentiated tends to lose proliferation ability. Cell density contributes in cell differentiation which in turn affects cell growth. Because of cell density limitation and physical restrictions imposed by organ culture geometry, most organ cultures do not grow or if they do proliferation is limited to the outer cell layers.

Procedure for Organ Culture

Growing a whole organ or fragment of organ is simple as cell cultures although certain conditions are need to be met. Culturing of an organ involves following steps:

- Selection of desired organ tissue.
- Dissection and collection of the organ tissue.
- Size of the tissue is reduced to less than 1mm in thickness as it contains sufficient numbers of cell.
- Place the organ tissue on a support such as raft, grid etc. at liquid- gas medium interface which allows exchange of gases and nutrients.
- Incubate the tissue with support in CO_2 incubator maintained at 37C.
- Change the medium on frequently basis as medium components tends to get depleted with passage of time.
- Examine the organ culture on routine basis for its morphological changes using microscopic technique. They further can be analyzed using histology, autoradiography and immunochemistry.

Techniques for Organ Culture

With the passage of time, various techniques have been developed to overcome the problems raised by cell culture methods. Following techniques has been employed for organ culture:

1. Plasma Clot or Watch Glass Method

- It was introduced by Fell and Robison. Prepare as plasma clot by mixing chicken plasma and chick embryo extract (50 %) in a ratio (3:1 drop).
- Explant is cultured by placing it on prepared plasma clot kept in watch glass.
- To minimize loss of clot by evaporation, 1-2 watch glasses are kept in Petri dishes piled with cotton wool or moist filter papers. Lid may or may not be closed and sealed with paraffin wax.
- Incubate Petri dish at 37.5°C and provide fresh plasma clots in every 2 to 4 days for avian and mammalian tissues.

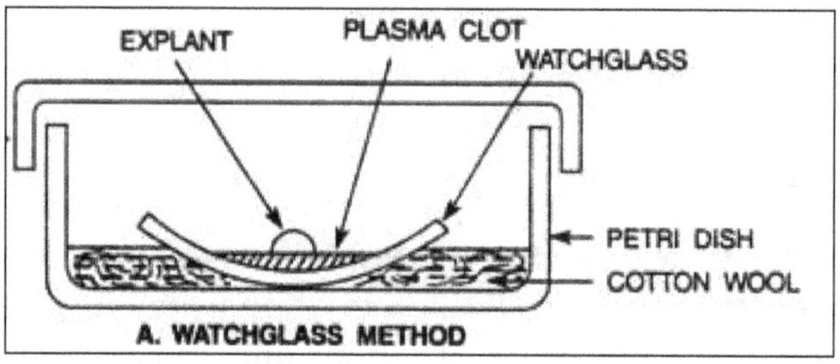

Figure: Watch glass/ clot method.

2. Raft Method

- This technique was introduced by Chen. Explants are placed on the raft of lens paper or rayon acetate strips, which allow floating on serum in watch glass. Silicone treatment is given to enhance floatability of lens paper/rayon acetate. The rayon acetate strip is preferable over lens paper because of its solubility in acetone that makes it advantageous during histological procedures. A single raft can occupy four or more explants.

3. Combination Method

- It combines method of watch glass and raft techniques. In this method first we place explants on the raft made up of suitable material, then it is placed on plasma clot. This overcomes the problem of sinking of explant in medium due to liquefaction of plasma clot.

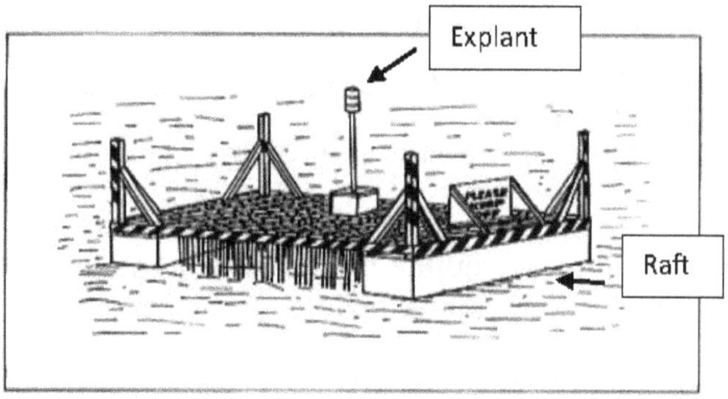

Figure: Raft method.

4. Agar Gel Method

- Agar gel (1%) is mixed with medium and poured into embryological watch glass. Explant is placed on surface of gelled medium and sealed with paraffin wax. Sub culturing is done every 5 to 7 days in fresh agar gels. Examine the explants for changes in growth pattern with stereoscopic microscope. The technique is used to study many developmental aspects of normal organs as well as tumors.

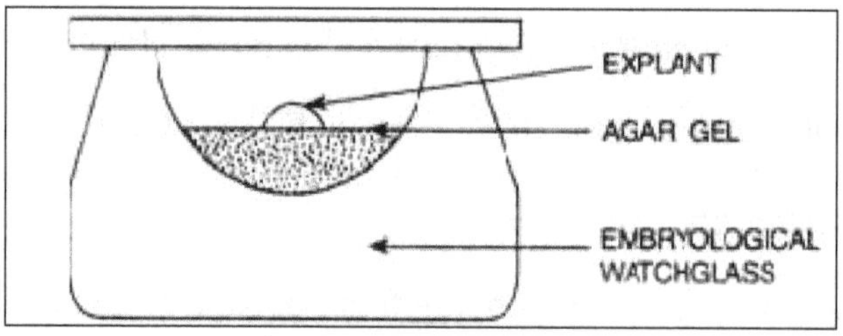

Figure: Agar gel method.

5. Grid Method

- It uses a grid (25 mm × 25 mm) made up of piece of wire mesh or perforated stainless steel sheet whose edges bent to form 4 legs of approx 4 mm height. Skeletal tissues are kept directly on grid while soften tissues are firstly placed on raft and then on grid. Latter is transfer to culture chamber filled with medium up to the level of grid. Supply of O_2/CO_2 is provided into chamber maintain constant level of O_2/CO_2 concentration. This method is best suited for studying growth and differentiation of adult as well as embryonic tissues.

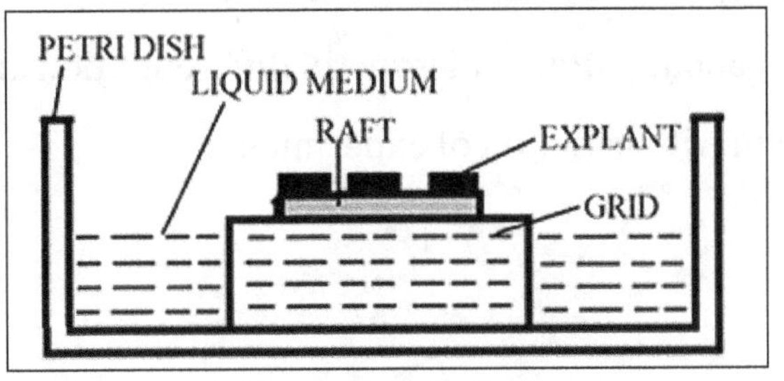

Figure: Grid Method.

Cyclic Exposure to Medium and Gas Phase

It is best suited technique for long term culture (4-5 months) of human adult tissues. Explants (2-8) are placed on culture dish and flushed with medium. Dishes are kept in chamber mounted on rocker platform containing suitable gas mixture for cyclic exposures of medium and gas.

Applications of Organ Culture

- It has been used to study morphogenesis, differentiation, and function in excised organs or presumptive organs.
- Used for comparison of growth and behavior of explanted organs.
- Effects of variables like radiation, several hormones, carcinogens and mutagens can be explained by means of organ culture.
- Mammary gland cultures using mouse model are commonly used to investigate responses of several hormones and its secretary actions.
- It is also used as a vehicle for mammary tumor agent.
- It can be used to examine toxic effects of steroid hormones on mammary adenocarcinomas of C3H mice.
- It has been used for in vitro production of genotypically mosaic embryos from fused eggs because organ culture allows cultivation of mouse ova.
- Organ culture can be used to explain controls of morphogenesis and embryonic induction.
- It has been also used for investigation of the effects of specific mesenchymal elements upon epithelial structures.
- It has been used to study the developmental anomalies (retinal dystrophy in C3H animals), inherited diseases and defects (kidney rudiments of prospective kidneyless (Sd/Sd) mice.

Advantage and Disadvantage of Organ Culture

Advantages

- It allows control of physicochemical & environmental factors like - pH, temperature, dissolved gases (O_2 and CO_2) and osmolarity.
- It gives better understanding of cell to cell interactions and hormonal control.
- Well suited for cytological and immunological studies.
- Provide information about patterns of growth, differentiation and development of tissue.
- Investigate a problem with lesser on of experiments.

Disadvantages

- Slow growth.
- Cost of expenditure is high on maintenance.
- It can be maintained only for a few months.
- Low reproducibility with high variability.
- Fresh explant is needed for every experiment.

Current Progress

- Anthony Atala made a successful trial for in vitro growth of bladder at Wake Forest Institute for Regenerative Medicine.
- Doris Taylor cultured a beating rat heart at the University of Minnesota.
- H. David Humes cultured an artificial kidney at the University of Michigan.
- A jawbone has been cultured at Columbia University.
- For production of heart tissue, scientist have successfully used silk cut from silkworm cocoons as growth scaffold.
- Regeneration of heart tissue is not possible if it gets damaged. The scientists have reported producing replacement patches for damaged part of heart is of great interest. The experiment has been carried out using rat heart cells for production of functional heart tissue. In order to further test applications to humans as a cure, a way to transform human stem cells into heart tissue would have to be found.
- In 2016, an experiment was carried out in which human cells were used to assemble intricately structured hearts. Although generated heart was structurally and functionally immature but proved we were only one step away from making a heart from stem cells.
- A study was carried out in 2017 by a group of scientists from Salk Institute for Biological Studies. They created a pig embryo whose segment of DNA (essential for organs growth)

was edited out. They then introduced human stem cells inside the pig embryo to have the human DNA fill in the gaps. They have got huge success in this regard.

Cell Cryopreservation

Protection of biological samples such as organs, tissues, extracellular matrix, cells, organelles, or any other biological entity from the damage of unregulated chemical kinetics by the process of cooling to very low temperature is called Cryopreservation. Cryopreservation typically done between -80°C to -196°C using solid carbon dioxide or liquid nitrogen. The freezing causes damage to the samples as formation of ice crystals or by increase in solute concentration. Conventionally, materials to be cryopreserved were coated with numbers of molecules prior to freezing called cryoprotectants such as glycerol. The development of cryoprotectant is going on as several cryoprotectants are toxic to the cells. It is very important to optimize the variables of cryopreservation which has direct effect on survival. The variables of cryoprotectant includes the nature of the cryoprotectant, concentration, temperature at which it added, cooling and warming rates, storage temperature, and rate at which the cryoprotectant is removed. Regardless of its tedious process optimization of cryoprotectant was markedly successful and various methods were developed for a range of cells such as haemopoietic cells, lymphocytes, erythrocytes, spermatozoa of several species, endocrine cells, and several strains of tissue culture cell.

Early successes of cryopreservation are well documented in the monograph of A.U. Smith "Biological Effects of Freezing and Supercooling". The early success of cryopreservation inspired the researchers to reveal the mechanisms. It is very important to study the total quantity of ice formed and their cellular location; another important variable are toxicity of cryoprotectant and their dependency on temperature i.e. level of toxicity in relation to temperature and also the changes in volume due to osmotic changes.

Natural Cryopreservation

Tardigrada (water-bears), a microscopic multicellular organism can survive very low temperature. It prevents crystallization of water by replacing its most of internal water with trehalose a sugar and hence prevents itself from damage of cell membrane. Dr. Kenneth B. Storey, a Canadian researcher discovered the incident of "freezing frog". The wood frog can tolerate and survive several freezing and thawing. Its blood and tissues are frozen during winter. Upon overwintering, wood frog can accumulate urea in tissues and convert glycogen into glucose when it sense internal ice formation. Both urea and glucose have the cryoprotectant properties and prevent osmotic shrinkage of cells. The organism can cease their life function during winter. These are some species of frogs (*Rana sylvatica, Hyla crucifer, Hyla chrysoscelis, Hyla versicolor, and Pseudacris triseriata*), one of salamanders (Hynobius keyserlingi), one of snakes (*Thamnophis sirtalis*) and three of turtles (*Chrysemys picta, Terrapene carolina, Terrapene ornata*).

Theoretical Background of Cell Freezing

The optimal cell freezing is characterized by formation of intracellular crystal is minimum and utmost cells are viable upon thawing. At high solute concentration foci formed are minimum. To accomplish optimum freezing following points are must be considered:

- The cooling of samples must be done slowly so that water can be escaped.
- Cooling should be fast enough to avoid formation of crystal.
- Cryoprotectants of hydrophobic nature should be used.
- Store the sample at ultra-low temperature.
- Thaw the samples rapidly to minimize crystal growth and solute gradients.

Methods of Cryopreservation

Slow Freezing Method

In slow-freezing, cells in a medium are cooled to below freezing point. In this method cells with cryoprotectant are cooled to below freezing point. At certain stage, water molecule converted in pure water crystal and unfrozen fraction of cells and their solutes are remains. Upon cryopreservation unfrozen fraction of sample decreases while concentration of salts, sugar and cryoprotectant increases. This osmotic change in cell cause outward movement of water. In slow freezing, to reduce the risk of intracellular ice formation it is necessary to efflux water from the cell. With the progression of freezing, viscosity of unfrozen fraction becomes very high and this unfrozen fraction converted to amorphous solid that contains no ice crystals.

Manual Freezing

In this method the temperature of cells/semen is decreased during adding of cryoprotectant step-by step and then the samples are immersed into liquid nitrogen. The chilling rate of the specimen from the room temperature to 5°C should be 0.5 to 1°C/min. Then the samples are frozen from 5°C to -80°C at the rate of 1-10°C/min and finally the specimen is plunged into liquid nitrogen at -196°C.

Slow Programmable Freezing

Firstly liquid nitrogen is poured into the tank and then the programmed through a machine with the help of a software and data logging in to gain the cooling from 20°C to −80°C at a rate of 1.5°C/min and then at 6°C/min; After completion of the freezing, straws are taken out for their storage in liquid nitrogen at −196°C. The whole process takes ~40 min. This technique is easy for using. It does not need continuous operator intervention. The process increases the reproducibility of the freezing operations.

Rapid Freezing Method

Ultra Rapid Freezing

It is the technique between the slow freezing and vitrification and faster than the slow freezing technique. It requires low concentration of cryoprotectant and do not need any programmable machines. The early cleavage stage embryo has been cryopreserved by using this method. The two-celled embryos were frozen followed by the exposure for 2 to 2.5 min. to high concentrations of 3 to 4 M dimethyl sulfoxide and 0.25 M of sucrose and plunged into liquid nitrogen. The embryos are then thawed in a waterbath at 37°C sealed in a plastic bag.

Vitrification

In vitrification samples are solidified to form a glass like structure and avoid the development of intracellular and extracellular ice. In vitrification process high concentration of cryoprotecting agents and/ or very high cooling rates are used to accomplish the cryopreservation. It is a cost effective as well as time effective process. It does not involve any expensive instruments and take quite a few minutes as compared to slow freezing.

Advantages of Vitrification

- No ice crystal formation is occurred during this process.
- Rapid equilibrium is achieved.
- Absence of water leak after equilibrium.
- A very short time period is required for the cryopreservation.
- Minimum damage of the membrane lipids will be occurred during this process of cryopreservation.
- This procedure of cryopreservation is very simple as compared to the other methods of cryopreservation.
- As compared to the other methods in this method do not require any equipment and machine.

Disadvantages of Vitrification

- This method of cryopreservation requires a well-established protocol.
- Ultra rapid thawing is required for this method of cryopreservation.
- Ice crystal formation may be occurred during hesitating thawing procedure.
- A test for the genetic damage is needed to carry out for the survivability and viability of the samples.

Factors Affecting Cryopreservation

- Water: Crystallization of water inside the cell increases the volume. It causes physical damage to the cell membrane and other cellular organelles which resulted in cell death.
- Cryoprotectant: Almost in every cell type, it is important to add cryoprotective agents to minimize the injury due to freezing and thawing. The effect of number of cryoprotective agent in cryopreservation has been extensively studied in mammalian cells e.g. erythrocytes and spermatozoa. The cryoprotectants must have the properties like membrane permeability, high osmolarity, non-crystallized material, easily super cooling and low melting point.

- Membrane permeability: The permeability of the cell membrane to the solution is an important factor in cryopreservation. The extent to which a cell can shrink and re-swell is depending on the permeability of cell membrane and on the concentration of cryoprotectant.
- Seeding temperature: The cell's viability is significantly affected by the ice seeding temperature. The intracellular ice formation interrelated with the extracellular ice-seeding temperature. The frozen storage of human oocytes, sheep and mouse cells are very sensitive to seeding temperature. An improper ice seeding temperature can leads to membrane destruction, cell dehydration and other fatal damage.
- Buffer: Buffers resists the change in pH during freezing, pKa value of a buffer closer to optimum freezing pH is more stable during temperature changes in cooling. Zwitterions are readily soluble in water but do not cross the biological membrane. Examples of zwitterions are BES (N,N-Bis-(2-hydroxyethyl)-2-aminoethanesulfonic acid), HEPES (N-(2-Hydroxyethyl)-piperazine-N'-ethanesulfonic acid), MES (2-(N-Morpholino)- ethanesulfonic acid), MOPS (3-(N-Morpholino)-propanesulfonic acid), PIPES (PiperazineN,N'-bis(2-ethanesulfonic acid)), TES (2-[Tris(hydroxymethyl)-methylamino]- ethanesulfonic acid) and Tricine (N-[Tris(hydroxymethyl)-methyl]-glycine).
- Salt concentration: High concentration of solutes in an unfrozen fraction affects the cell's stability. The loss in cell membrane stability and protein denaturation are resulted when a cell encounters dehydration and high salt concentration. Extracellular salts may enter into the cell when they are present in excessive concentration. High salt concentration also cause efflux of water and resulted in the decrease in volume of the cell by more than 50% of its original volume and leads to structural deformation of the cell.
- Cooling rate: It is commonly observed in every biological system and cells. It has a specific optimal cooling rate. Bellow or above this rate the survival of the cell is decreased during the slow cooling damage or fast cooling damage. Different types of cells and different types of biological materials exhibit different cooling rates. The volume to surface area ratio of the cells and biological systems such as embryos or pieces of tissue determine the optimum cooling rate. It also affects the cells membrane permeability to the water and cryoprotectants.
- Cryodamage: It is the damage of the cells, tissues and biological entity due to freezing at low temperature. There are two types of damages occurs during freezing of samples, one is physical damage and the other is chemical damage. The physical damage is caused by rapid dehydration and hydration, sudden fall in temperature and shrinkage of cell. The chemical damage is caused by the cryoprotectants. The cryoprotectant causes toxicity to the samples, increase concentration of free tubulin, reduction in size of spindle, disorganization of the microtubule, induces chromosomal abnormality also has detrimental effects on cellular organelles such as golgi body, endoplasmic reticulum, mitochondria, cytosol etc.

Applications of Cryopreservation

- In the preservation of the animal cells, tissues and organs etc.
- In prevention of endangered species of animals.
- Genome resource banking, in assisting reproductive technologies, preservation of embryo, oocytes, ovarian tissue, semen, testis tissues etc.

Biotechnology and Conservation

The first recombinant DNA experiment was conducted in the year 1973 and also was the start of the modern biotechnology. The commercialization and the real application of the recombinant DNA technology was started in the year 1980, when the Supreme Court of the United States has given the permission to patent the genetically modified microorganism. From the last two decades, lot of work has been carried out on using biotechnological tools and approaches on bioprospecting of biological wealth throughout the world. The main objectives are:

- To characterize and the prospecting of genes and bio-molecules.
- For the conservation of the biodiversity of various eco-geographical regions of the world.

By using the biotechnological approached conservation strategies has been developed for the endangered animal species. The gene prospecting studies have also been carried out from the last one decade in search of different stress tolerant genes, these genes has been characterized and cloned to raise the transgenic animals from these genes. Biotechnology can play an important role in ex-situ an in-situ conservation of animals.

Advantages and Disadvantages of Animal Cell Culture

Advantages of Animal Cell Culture

- Physiochemical and physiological condition: Role and effect of pH, temperature, O_2/CO_2 concentration, and osmotic pressure of the culture media can be altered to study their effects on the cell culture.
- Metabolism of cell: To study cell metabolism and investigate the physiology and biochemistry of cells.
- Cytotoxic assay: Effect of various compounds or drugs on specific cell types such as liver cells can be studied.
- Homogenous cultures: These cultures help study the biology and origin of the cells.
- Valuable biological data from large-scale cell cultures: Specific proteins can be synthesized in large quantities from genetically modified cells in large-scale cultures.
- Consistency of results: Reproducibility of the results that can be obtained by the use of a single type/clonal population.
- Identification of cell type: Specific cell types can be detected by the presence of markers such as molecules or by karyotyping.
- Ethics: Ethical, moral, and legal questions for utilizing animals in experiments can be avoided.

Disadvantages of Animal Cell Culture

- Expenditure and expertise: This is a specialized technique that requires aseptic conditions, trained personnel, and costly equipment.
- Dedifferentiation: Cell characteristics can change after a period of continuous growth of cells in cultures, leading to differentiated properties compared to the original strain.
- Low amount of product: The miniscule amount of mAB and recombinant protein produced followed by downstream processing for extracting pure products increases expenses tremendously.
- Contamination: Mycoplasma and viral infection are difficult to detect and are highly contagious.
- Instability: Aneuploidy chromosomal constitution in continuous cell lines leads to instability.

Applications of Animal Cell Culture

The nine applications of animal cell culture are:

Model Systems

Cell cultures provide a good model system for studying:

- Basic cell biology and biochemistry.
- The interactions between disease-causing agents and cells.
- The effects of drugs on cells.
- The process and triggers for aging.
- Nutritional studies.

Toxicity Testing

Cultured cells are widely used alone or in conjunction with animal tests to study the effects of new drugs, cosmetics and chemicals on survival and growth in a wide variety of cell types. Especially important are liver- and kidney-derived cell cultures.

Cancer Research

Since both normal cells and cancer cells can be grown in culture, the basic differences between them can be closely studied. In addition, it is possible, by the use of chemicals, viruses and radiation, to convert normal cultured cells to cancer causing cells. Thus, the mechanisms that cause the change can be studied. Cultured cancer cells also serve as a test system to determine suitable drugs and methods for selectively destroying types of cancer.

Virology

One of the earliest and major uses of cell culture is the replication of viruses in cell cultures (in place of animals) for use in vaccine production. Cell cultures are also widely used in the clinical detection and isolation of viruses, as well as basic research into how they grow and infect organisms.

Cell-based Manufacturing

While cultured cells can be used to produce many important products, three areas are generating the most interest. The first is the large-scale production of viruses for use in vaccine production. These include vaccines for polio, rabies, chicken pox, hepatitis B and measles.

The second is the large-scale production of cells that have been genetically engineered to produce proteins that have medicinal or commercial value. These include monoclonal antibodies, insulin, hormones, etc. The third is the use of cells as replacement tissues and organs. Artificial skin for use in treating burns and ulcers is the first commercially available product. However, testing is underway on artificial organs such as pancreas, liver and kidney.

A potential supply of replacement cells and tissues may come out of work currently being done with both embryonic and adult stem cells. These are cells that have the potential to differentiate into a variety of different cell types. It is hoped that learning how to control the development of these cells may offer new treatment approaches for a wide variety of medical conditions.

Genetic Counselling

Amniocentesis, a diagnostic technique that enables doctors to remove and culture fetal cells from pregnant women, has given doctors an important tool for the early diagnosis of fetal disorders. These cells can then be examined for abnormalities in their chromosomes and genes using karyotyping, chromosome painting and other molecular techniques.

Genetic Engineering

The ability to transfect or reprogram cultured cells with new genetic material (DNA and genes) has provided a major tool to molecular biologists wishing to study the cellular effects of the expression of these genes (new proteins). These techniques can also be used to produce these new proteins in large quantity in cultured cells for further study. Insect cells are widely used as miniature cells factories to express substantial quantities of proteins that they manufacture after being infected with genetically engineered baculoviruses.

Gene Therapy

The ability to genetically engineer cells has also led to their use for gene therapy. Cells can be removed from a patient lacking a functional gene and the missing or damaged gene can then be replaced. The cells can be grown for a while in culture and then replaced into the patient. An alternative approach is to place the missing gene into a viral vector and then "infect" the patient with the virus in the hope that the missing gene will then be expressed in the patient's cells.

Drug Screening and Development

Cell-based assays have become increasingly important for the pharmaceutical industry, not just for cytotoxicity testing but also for high throughput screening of compounds that may have potential use as drugs. Originally, these cell culture tests were done in 96 well plates, but increasing use is now being made of 384 and 1536 well plates.

Ethical Issues in Animal Cell Culture

Ethics is a system of moral principles. Ethics is a problem or situation that requires a person or organization to choose between alternatives that must be evaluated as right (ethical) or wrong (unethical). Historically medical ethics is very old may be traced to guidelines on duty of physicians such as Hippocratic oath Working with human and animal tissue presents a number of ethical problems involving procurement, subsequent handling and ultimate use of material.

Ethical problems involve:

- Conflict.
- Choosing between equally desirable or undesirable alternatives.
- Balancing options.

Ethics is concerned about what should be considered right and wrong.

Ethics are considered as:

- Ethics are moral principles.
- Ethics are the difference between good and bad.
- Ethics are the difference between right and wrong.
- Ethics are based on value system.
- Ethical norms are not universal – depends on sub culture of the society.

Ethics are not considered as:

- Ethics are not the same as feelings of person.
- Ethics are not according to religion.
- Ethics are not following the law of countries.
- Ethics are not following culturally accepted norms.
- In facts ethics are not science.

Broad Categories of Ethical Issues

The ethical issues may be broadly associated with transgenic animals, mammalian cloning and animal tissues.

- First are issues that pertain to the impact of this technology on the animals themselves.
- Second are issues that relate to the institutions and procedures that govern the research and applications context within the agri-food system.
- Finally, there are issues that relate to the relationship between humans and other animals.
- Ethically uses of animal tissues.

Purpose and Scope

- The main purpose and scope of the bioethics mainly deal with human inquiry during the experiments.
- Bioethics deals with the ranging from debates over boundaries of life like abortion, euthanasia, surrogacy, allocation of limited health care resources (e.g. organ donation).
- Bioethics deals with right to refuse medical care for the religious or cultural reasons if any.
- Bioethicists often disagree among themselves over precise limits of their discipline on certain issues.
- Bioethics deals with whether field should concern itself with ethical evaluation of all questions involving biology and medicine or only a subset of these questions.
- There are two basic groups of the bioethicists worldwide.
- Some bioethicists would have narrow ethical thought & evaluation only to morality of medical treatments or technological innovations and timing of medical treatment of humans.
- Others would consider broaden scope of ethical evaluation to include the morality of all actions that might help or harm organisms capable of feeling fear.

Principles of Bioethics

- Beneficence (or Paternalism): Physician-centered decisions.
- Non-maleficence: First do no harm.
- Autonomy: Patient's right to self-determination.
- Justice: Fairness.

Principle of Beneficence (Related to Benevolence)

- Ethical one should cause to be positive assistance to others means one person should help ever hurt never the other person.

- Ethically every person should abstain from harm (minimalist principle of non-malfeasance) by helping them to achieve benefits which will further their important and legitimate interests.
- Harm principle as a form of beneficence.
- It is moral or ethical duty of every person to prevent an individual from harming another person.
- Ethically one should give to persons best what they are pay for them.
- Ethically one should give what they deserve.
- Ethically one should give what they can legitimately claim.
- Treating equals equally unless there is a morally relevant difference requiring persons to be treated unequally/differently.
- Consideration must often be given to a proper allocation of benefits and burdens within the social context.

Principle of Non-maleficence

There are some situations in which some type of harm seems inevitable and we are usually morally bound to choose the lesser of the two evils, although the lesser of evils may be determined by the circumstances. For example, most would be willing to experience some pain if the procedure in question would prolong life. However, in other cases, such as the case of the patient dying of painful intestinal carcinoma, the patient might choose to forego CPR in the event of a cardiac or respiratory arrest. The reason for such a choice is based on the belief of the patient that prolonged living with a painful and debilitating condition is worse than death. It is also important to note in this case that this determination was made by the patient, who alone is the authority on the interpretation of the "greater" or "lesser" harm for him.

There is another category of cases that is also confusing since a single action may have two effects, one that is considered a good effect, the other a bad effect. How does our duty to the principle of non-maleficence direct us in such cases? The formal name for the principle governing this category of cases is usually called the principle of double effect. A typical example might be the question as to how to best treat a pregnant woman newly diagnosed with cancer of the uterus. The usual treatment, removal of the uterus is considered a lifesaving treatment. However, this procedure would result in the death of the foetus. What action is morally allowable, or, what is our duty? It is argued in this case that the woman has the right to self-defence and the action of the hysterectomy is aimed at preserving her life. The unintended consequence (though undesired) is the death of the foetus. There are four conditions that usually apply to the principle of double effect:

- Action itself must not be intrinsically wrong; it must be a good or neutral act.
- Only good effect must be intended, not bad effect, even though it is foreseen.
- Bad effect must not be means of the good effect.
- Good effect must outweigh the evil that is permitted.

Other problems arise when the primary patient cannot decide for himself and others must determine what is in the best interest of the patient or what constitutes the lesser harm. In Washington State, the law actually guides the surrogate to offer "substituted judgment" if known, or to follow the course of action that will serve the "best interests" of the patient as determined by reasonable judgment.

Principle of Autonomy

- Ethically a person should be free to perform whatever action he/she wishes.
- Ethically regardless of risks as perceived by others should be benefited.
- Ethically provided it does not impinge on the autonomy of other persons.

Principle of Justice

- The principle of justice could be described as the moral obligation to act based on fair adjudication between competing claims.
- Principle of justice is linked to fairness, entitlement and equality.
- Principle of justice in health care ethics can be subdivided into three categories:
    - Fair distribution of scarce resources (distributive justice).
    - Respect for people rights (rights-based justice).
    - Respect for morally acceptable laws.

Criticism of Bio-ethics

- There are many reasons to criticise the bio-ethics policy in our country or many other parts of world.
- Bioethics tends to focus its attention on problems that arise from "too much care" for patients in industrialized nations, while giving little or no attention to ethical problem of too little care for the poor.
- Farmer characterizes bioethics of handling morally difficult clinical situations, normally in hospitals in industrialized countries, as "quandary ethics".
- He does not regard quandary ethics and clinical bioethics as unimportant; he argues, rather, that bioethics must be balanced and give due weight to poor.
- Bioethics has been condemned for its lack of diversity in thought, particularly with regards to race.
- Even as field has grown to include areas of public opinion, policymaking, and medical decisions, little to no academic writing has been authored concerning the intersection between race-especially the cultural values fill in that construct and bioethical literature.

References

- Culture-conditions-and-types-of-growth-media-for-mammalian-cells, biomedical-tissue-culture: intechopen.com, Retrieved 02, June 2020
- Primary-cell-culture-3-techniques-with-diagram, primary-cell-culture, cell-10527: biologydiscussion.com, Retrieved 15, Feb 2020
- Types-of-subculture-of-cell-2-types-with-diagram, cell-10514: biologydiscussion.com, Retrieved 05, March 2020
- Organ-culture-and-its-techniques-animal-biotechnology, animal-biotechnology-13273: notesonzoology.com, Retrieved 25, May 2020
- Applications-of-animal-cell-culture, animal-cell-culture, cell: biologydiscussion.com, Retrieved 02, July 2020

3

Transgenic Animal Technology

The methods used to incorporate foreign genes into the genome of an animal so that these genes are inherited and expressed in the offspring fall under transgenic animal technology. Transgenic animals are those animals whose genome has been modified to have desired traits. This chapter elaborates methods of creation of transgenic animals, and applications and ethical concerns of transgenic technology.

"Transgenic technology", also known as transgenesis, is a process of integration of exogenous genes or cloned DNA into genetic material of an organism producing DNA alterations to give permanent change of effect. Organisms containing integrated sequences of DNA are called "Transgenic Animal" or "Genetically Modified Organisms (GMO)". Foreign genes which insert itself into the germ line of the animal and transmitted to the progeny are called as "transgenes". Transgenesis may involve in vivo alteration of body function of an organism, rather than individual cells. So, the ability of transgenesis to introduce genes into eggs or early embryos has given an extension to embryological study providing scope for the production of farm animals of desirable traits. Therefore, transgenesis had been proved to be far better and complex than cell-culture based applications by imparting knowledge at embryological level. Transgenic animals are used for variety of purposes such as in research areas, to produce valuable drugs for animals/human welfare, to generate animal models and cell lines for studying pathogenesis and/or prevention of disease in humans, to study genetic regulation of development and growth, to produce conventional products in more quantity (e.g. meat, milk, leather, wool etc.) and for the synthesis of non-conventional products e.g. Interferon, α- antitrypsin etc.

This sector of genetic engineering has raised a great public concern, equal in both consequences and potential including the outlined basic research technologies toward fruitful applications. The development and use of transgenic organisms involve many complex surrounding issues. In addition, in higher organisms many scientific and technical problems are associated with genetic engineering that is often difficult to handle. The problem rise due to genome complexity and difference in development pattern of plants and animal. Despite of all these hurdles, methods for transgenics are now well established. In this module, considering basic concept of transgenic animals, we will learn how to produce transgenic animals, their applications in medicine, agriculture and industrials fields and their ethical, social and legal concerns.

Methods of Creation of Transgenic Animals

Gene/Transgene Transfer Methods

The term transformation is used in animals to describe phenotypic alteration in cells. Therefore, Transfection, also called gene transfer methods, is a procedure that introduces foreign genes (transgenes) into living organism to produce genetically modified organisms or transgenic animals.

In last few decades, a wide repertoire of gene transfer techniques has been evolved in field of transgensis. One of which is physical transfection. It enables the direct transfer of nucleic acids (DNA/RNA) by physical or mechanical means into the cytoplasm, or nucleus without the usage of foreign substances like lipids. It is most preferable method for gene transfer as it avoids the complications associated with viral and chemical strategies.

DNA Microinjection

Microinjection is a useful technique that plays an important part in the generation of genetically modified mouse and rat models. Different microinjection systems can exist, the construction and configuration of which directly depends on the type of experiment.

DNA microinjection is the dominating technique leading to random integration of a transgene via the introduction of DNA into the pronucleus of a developing zygote. Following fertilization of a mouse egg, the male and female pronuclei remain separated for a few hours before they fuse to make the zygotic nucleus, thus allowing for microinjection of the desired genes into the larger male pronucleus. Eggs that can survive the injections are transferred into the oviducts of a foster mother i.e., pseudopregnant female mice for the generation of Founder mouse, from which permanent transgenic lines can be established. The presence of transgenes in the offsprings can be identified by PCR analysis or Southern blot hybridization.

Two kinds of DNA injections are included:

- Plasmid Vectors
- BAC Injections

Features of DNA Microinjection

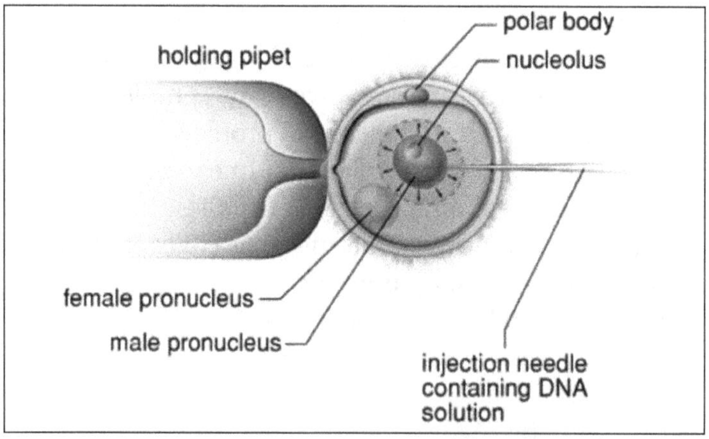

Figure: Diagram of DNA Microinjection into a Pronucleus.

- The amount of DNA delivered per cell is not limited and the delivery is precise, both of which increase the chance for integrative transformation.
- As a direct physical approach, this technique is host-range independent.
- Cells that are injected with the DNA material have high regeneration potential.

- Although pronuclear microinjection is popular, this method for transgenesis has some disadvantages such as low success rate and mosaic founders.

This technique not only provides more consistent gene expression between animals but also requires less effort. Moreover, fewer animals are needed to perform experiments using this method.

ES Cell Injection and CRISPR Injection

Before producing the desired genetically modified mouse models, the embryonic stem (ES) cell must be transferred into blastocysts first to create chimeras. Similarly, CRISPR injections are needed for modification of the endogenous gene locus.

Embryonic Stem Cell-mediated Gene Transfer

The term embryonic stem cell (ESC) was first used in 1981 to denote a cell line isolated directly from mouse embryo. ESC are harvested from inner cell mass (ICM) of early mouse embryos from blastula stage. They have the potential to participate in embryogenesis with high efficiency when combined with normal embryos to form chimaeras. Their efficiency is maintained even after they are altered or manipulated for gene transfer and can retain their full potential when grown in culture to differentiate into all cells type. Due to this reason, it provides an excellent vehicle for gene transfer into living organism. This is known as ESC mediated gene transfer.

This method is most common in mice. Initially, blastocysts are isolated from blastula stage from mouse embryo. Inner cell mass (ICM) is removed from blastocysts and cultured with embryonic stem cells. DNA molecules (transgene) containing desired structural gene is constructed by using recombinant DNA method with promoter and enhancer sequences.

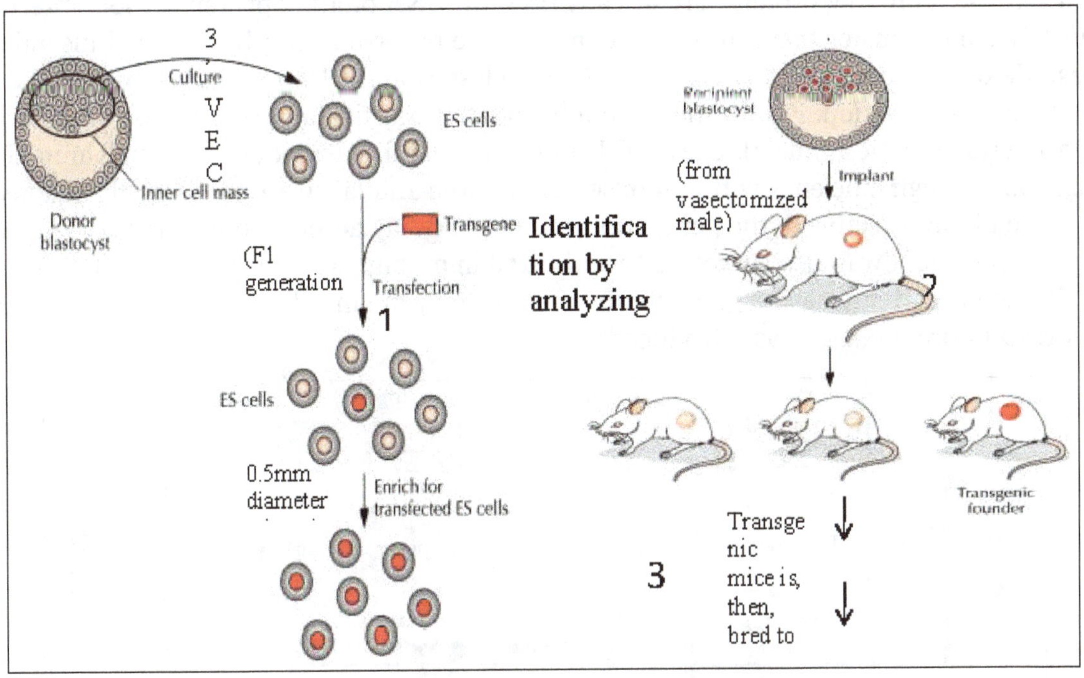

Figure: Embryonic stem cell mediated transfer.

ES cells are transformed by exposing cultured cells to the foreign DNA. Like pronuclear microinjection, introduction of DNA into ES cells is also random. If the injected DNA is having similar sequence to part of the mouse genome, it may undergo "homologous recombination" and integrate itself at a specific site as a single copy. Successfully transformed ES cells are then selected and injected into cavity of host blastocysts of different mouse strain and re-implanted into uterus of foster mother (pseudo-pregnant mother).

After the birth of offspring, they are tested for desirable gene by examining their small part of tissue obtained from tail. An ES cell colonizes host embryo and contributes to the germ line. If ES cells have contributed to their germ cells, atleast 10 – 20% will show positive results with heterozygous condition in F1 generation. To establish strain, homozygous condition is produced by mating two heterozygous offsprings (F2 generation). Mating of heterozygous produces transgenic animal.

Embryonic stem-cell mediated gene transfer is time and cost intensive method but holds great importance. It helps in studying genetic control during developmental process by allowing the detection of transgenesis at embryonic level. Due to this, mouse genome can also be modified at embryonic level by inserting, removing or modifying DNA sequences, which could lead to Knock-out, knock-in or conditional mutant mice.

Sperm-mediated Gene Transfer

Sperm-mediated gene transfer (SMGT) was developed as an alternative technique for the production of transgenic animals. Sperm-mediated gene transfer (SMGT) is a procedure through which new genetic traits are introduced in animals by exploiting the ability of spermatozoa to take up exogenous DNA molecules and deliver them to oocytes at fertilization.

This ability allows sperm cell to bind exogenous DNA molecules in the subacrosomal region of its head, where DNA binding proteins are already present. DNA binding proteins, then, internalises foreign DNA and transfer them into oocyte at the time of fertilization by artificial insemination process. Major histocompatibility complex (MHC) class II and CD4 molecules are also involved in sperm/DNA interaction. Fertilization could be in vitro or artificial insemination or waterborne fertilization (for aquatic animals). The SMGT procedure was first introduced in small animal model, and reported high efficiency within mouse. Later, large animals were successfully adapted and found to be highly effective in the generation of human decay accelerating factor (HDAF) transgenic pig cell lines. It was also successfully optimized in mutagenic transgenic pigs which contain three different reporter genes i.e. green fluorescent protein, enhanced blue fluorescent protein and red fluorescent protein, and was introduced.

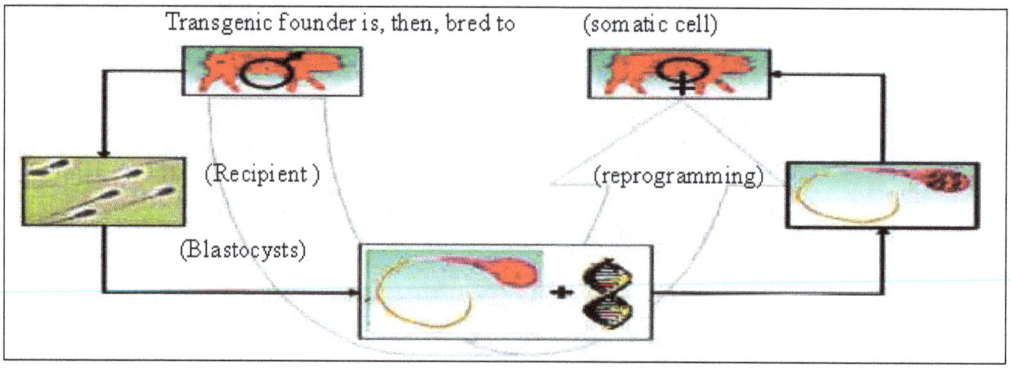

Figure: Sperm mediated gene transfer in pigs.

SMGT technique offer advantages of inexpensiveness and ease of use and high efficiency (5 – 60 %) in contrast to 0.5 – 4 % in microinjection. It also does not require embryo handling or expensive equipments. It provides broad applicability within species ranging from sea urchin to cattle.

Retrovirus-mediated Gene Transfer

Retroviral methodology is an effectual method to integrate interested genes into animal genomes. It is considered as a naturally occurring gene transfering system to integrate DNA into mammalian cells. In 1974, the first healthy mice, that carry copies of foreign viral DNA was made by microinjecting SV40 DNA into the blastocysts. Consequently, the germline transmission of viral DNA into the animal genome could be achieved by directly infecting preimplantation stage embryos with Moloney leukemia viruses (MLV). Therefore, the two types of retroviral vectors are established for creating transgenic animals; the first vectors can be derived from the genome of prototypic retrovirus (such as MLV) and the second vectors are derived from the genome of more complex retroviruses (such as lentiviruses). These virus holds the structural gene such as gag,pol and env which can be responsible for virulence characteristics of the virus. For safety reasons these structural genes of retroviruses are frequently modified by removing this part. Though, retroviral lines that used in transgenic animal experiments are ecotropic, they infect only mice or rats. This could be dangerous for human rather than rodent if correct precautions are not taken.

Retroviral vectors transgenesis was fruitful in many animal species, including livestock although the technique has some serious drawbacks. The first one is the transgenes brought by a retroviral vector cannot be expressed in the transgenic animal. Silencing of the transgene has been revealed to be affected by the enrollment of gene suppression machinery in the host cells by the promoter and enhancer sequences of the retroviral long-terminal-repeats (LTRs) and the following hypermethylation of the viral promoter sequences by de novo DNA methylation. Second, it is difficult to package a transgene larger than 10 kb due to the smaller size capacity of such viruses, which can seriously act to limits its use for delivering large genes. Third, it was indicated that the retroviral long-terminal-repeats (LTRs) can hamper the mammalian promoters to overcome or directing the host gene expression wrongly. the last is the breakdown of the nuclear envelope throughout mitosis is necessitates for infection of early embryos and yet the viruses over and over again show a postponement in transgene integration, subsequently germline transmission of the transgene was very low in which mosaic animals were produced.

Replication defective adenoviruses (DNA viruses) were used as vectors to overcome the limitation set by retroviruses in infecting non-dividing cells, rats and cattle. While transgenic mice, with a tendency for germline transmission were produced, similar success has not been shown in other species. The real innovation in the application of viral vectors in making transgenic animals came after the development of lentiviral vectors and their following use in creating transgenic animals. For the reason that of their capability for infecting a large spectrum of cells and embryo, lentiviral transgenesis is now a day the most efficient approach for making transgenic animals, including poultry and livestock.

Livestock species from which transgenic animals were made by a lentiviral vector-based method is developing. It was informed recently that transgenic pigs could be produced with high efficiency using lentiviral vectors, 13 and 31%. Moreover, it was exposed that more than 90% of the genetically modified animals expressed the transgene at a high level. This methodology was also effective in cattle. All of the resulting calves from the transgenic positive embryos expressed the transgene when lentiviral vectors carrying transgenes were used for infjecting oocytes before fertilization however; no transgenic calves were born when preimplantation stage bovine embryos were infected with the same lentiviral vector.

Regardless of the advantages in using lentiviral vectors, this technique has limitations. Alike to conventional prototypic retroviral vectors lentiviral vectors still have the limitation of about 10 kb of transgene DNA. The application of lentiviral vectors success in certain species seems not to secure the success of the other species. In spite of its achievement in mouse, rat, pig, cattle and chicken, no transgenic monkeys were made by lentiviral transfer. Another drawback, which is common to all transgenic approaches that encompass random insertion into an endogenous chromosome, it is possible for mutating an endogenous genome.

Somatic Cell Nuclear Transfer

Nuclear transfer commonly represented to as cloning that involves the insertion of donor nuclei took from either differentiated adult cells or stem cells into enucleated egg. The enucleated oocyte becomes the host for a nucleus that is transplanted from another cell, such as a skin cell. In this manner of reprogramming future development in which the enucleated oocytes are then transferred to a foster mother for maintenance of gestation or used to generate embryonic stem cells with a genetic match to the nucleus donor for therapeutic cloning. The first mammalian nuclear transfer experiments were demonstrated in mice in early 1980s,. The application of Somatic cell nuclear transfer or cloning techniques may have a great potential to get higher number of offsprings from a single female. Meanwhile the famous cloned sheep "Dolly" was born. Somatic cell nuclear transfer technology has become another methodology available for the production of transgenic animals. Cloning by nuclear transfer from adult somatic cells is an outstanding manifestation of developmental manipulability of the oocyt by placing nucleues in the cytoplasm that can alter the chromatin structure and control the development of oocyte. In SCNT, all of the donor cell's genetic information is not transferred, as the donor cell's left behind its own mitochondrial DNA in the original cell. The resultant hybrid cells utilize the mitochondrial structures of the egg in which it inoculated in. as a result the cloned sheep Dolly that are born from SCNT are not absolute duplicates of the donor of the nucleus. Yet it has a great benefit in transgenesis, rescuing endanger animal and gene therapy.

SCNT, in transgenesis is feasibly the one that has promoted the most increasing the efficiency and reducing costs. It can make genetically altered animals through the cloning of transgenic founder animals or transfection of nuclei with vectors of DNA expression. In this method the exploitation of the transfection of exogenous DNA in the nuclei is randomly assimilated into the genome using selective pressure. Likewise, the transgenic cells can be absolutely characterized with respect to the number of copies integrated, integration region and integration of the transgene prior to the nuclear transfer step. Even though the capability for development of the reconstructed embryos is lower, the majority of the offspring are transgenic which makes this technology much more efficient than pronuclear microinjection. Somatic cell nuclear transfer is applied in the gene therapy by using gene targeting method some successes are there. The first production of gene- targeted sheep by somatic cells nuclear transfer that encodes the recombinant human a1-antitrypsin (rhAAT) was efficient and reproducible. In cattle, mastitis and bovine spongiform encephalopathy free cattles were produced. Pigs that can produce synthesized polyunsaturated fatty acids produced Lai et al. The aother application of somatic nuclear transfer is interspecies nuclear transfer is a used to enable the rescuing of extinct animal, or even to bring back species after their extermination. The cloning of highly endangered or extinct species involves the use of a unique method of cloning. Interspecies nuclear transfer operates with a recipient and donor of two different organisms which are closely related species and within the same genus. The Gaur (*Bos gaurus*) nucleus combining successfully with a domestic cow (*Bos taurus*) can be a good example in this scheme.

The negative side of this technique is the capability to come into the reproductive system after targeting modification, screening, and amplification. Somatic cells of domestic animals have a limited lifespan in vitro for cloning, while some of them maintain totipotency after gene targeting. These aged cells reduce the efficacy of this method, causing in a high incidence of aberrations in cloned embryos, fetuses, and offspring. For instant McCreath et al produced live targeted sheep at an efficiency of less than 4% and Lia et al produced live targeted pigs at an efficiency of less than 2%. Furthermore, in many species, especially large farm animals, ES cells have not been successfully established. The improvement and solving of these problems could permit the increasing of practical application of SCNT to the generation of transgenic animals for different agricultural, biomedical, and veterinary purposes.

RNA Interference Mediated Gene Knockdown

RNA interference (RNAi) is a natural post transcriptional gene regulatory process found in all organisms including animals, plants and fungi is the silencing of specific gene expression facilitated by the accumulation of double-stranded RNA that marks in the suppression of gene expression by degrading mRNA. Therefore, RNAi can achieve the spatiotemporality and reversibility of gene expression modulation.

Gene silencing become an essential methods of regulating gene expression. It is difficult to remove an existing gene from an animal genome than inserting a foreign gene, unless a knock-out technique is used. This gene knock-out is, challenging, complex and irreversible, if the gene is once knocked out, it cannot be recovered again. The development of RNA interference (RNAi) technology partially overcome this problem, through cleavage of the expressed mRNA or blocking of gene expression, it allows specific, partial and reversible knock-down of the desired gene besides can achieve spatio-temporal regulation of gene expression. Synthetic siRNAs can be transfected into cells or early embryos for transient gene knockdown,. For stable gene suppression, the siRNA sequences must be integrated into a gene construct and constitutively expressed. In bovine embryos, siRNA mediated knockdown has been undertaken to suppress the prion protein (PRNP) gene. Porcine endogenous retrovirus (PERV) has been demonstrated in porcine primary cells using RNAi knockdown and in cloned piglets. In the case of fish , microinjected myostatin small interference RNA (siRNA) into fertilized eggs of zebra fish, which can eliminating myostatin mRNA and reducing the inhibitory effect of myostatin on muscle growth and generating muscle hyper growth fish.

The most important issue in the application of RNAi is how to design effective RNAi sequence and produce constitutively stable expression in cells since all of siRNA do not have an inhibitory effect equally some of them do have such characters. Although the tool of RNAi is not well understood and many enzymes proteins and that are comprised an undefined function, this technique is still promising as the combination of siRNA and lentiviral vector technology offers a method for highly efficient targeted gene knock- down in farm animals. It could be easily joined into existing breeding programs and it could be applicable in studies of disease therapy and genetic function analysis.

Zinc-finger Nuclease Gene Targeting Technique

The beginning of the zinc-finger nuclease (ZFN) technique, suggest a qualitative leap in gene targeting techniques. ZFN is involved one DNA binding domain and one non-specific endonuclease domain that bind and cut DNA at specific sites by introduce double-stranded DNA fragment. This extrinsic DNA transferred by the induction of the endogenous DNA repair procedure in which homology-directed repair or non-homology terminal junction then the modification of the cellular endogenous gene take place by using Highly specific endonucleolytic enzymes to attain, such as the Fok1 endonuclease cleavage domain bonded with zinc-finger DNA binding domains and custom-designed homing endonucleases collectively known as "molecular scissors". This technique overcomes the limitation of gene targeting efficiency, which is enhanced by five orders of magnitude. Genetically distorted zebra fish have been made by targeted gene inactivation directly in the zygote using mRNA coding zinc-finger nucleases. Gene knock-out rats and pigs have been produced by this approach.

Targeted genetically modified animals were obtain in less than four months using the ZFN technique which can be able reduced the time of experiment with high the efficiency. Microinjection of constructed zinc-finger nucleases (ZFNs) in embryos of mouse and rat was used to generate gene knockouts by introducing non homologous end joining (NHEJ)-mediated deletions or insertions at the target site. Using ZFN technology in embryos to introduce sequence-specific modifications (knock-ins) by means of homologous recombination in Sprague–Dawley and Long-Evans hooded rats and FVB mice, enables precise genome engineering to generate modifications such as point mutations, accurate insertions and deletions, and conditional knockouts and knock-ins. Zinc-finger nucleases (ZFNs) are powerful technolgy for producing gene knockouts (KOs) with high efficiency and it has a possibility to apply to other species. Whereas the demonstration of ZFN-mediated gene disruption has been used in expermental animals such as rats and mice, ZFNs have not been used to disrupt an endogenous gene of large domestic species. In recent times, cloned pigs that carries a biallelic ZFN brought the swine a1,3-galactosyltransferase (GGTA1) gene knockout. This approach unlocks a new era for the creation of gene knockout pigs, which could important both agriculture and biomedicine.

The ZFN-gene targeting technique significantly improves the capacity of precise genetic modification of the animal and the efficiency of gene integration, and allows the establishment of large animal models for human diseases and xeno-transplantations. The research of gene function and agricultural breeding can be also beneficial by this approach since it has the ability to target a specific site. However, there are some imperfections in the ZFN as a result of its initial stage in progress.

Transgenic Animals

A transgenic animal is one whose genome has been altered by the transfer of a gene or genes from another species or breed. Transgenic animals are routinely used in the laboratory as models in biomedical research.

Transgenic Mice

Transgenic mice are mouse models that have had their genomes altered for the purpose of studying gene functions:

1. Gene knockout: Gene knockout (KO) is the genetic technique comprised of the biotechnological tools which are used to make the gene inoperative or inactivate in the whole body of the organism by permanently changing or mutating the gene of interest in the organism. This can only be achieved at embryonic stage. Homologous recombination is used to swap over the normal copy of an exon with a mutated version of the DNA. The genes can be made non-functional by deleting or replacing or mutating the coding sequences of selected genes. There is no need to remove the complete gene from the organism. The mutated or altered gene sequences translate a truncated and non-functional protein. As a result, the gene is no more functional and has been completely omitted from the organism.

2. Gene targeting/knockdown: Gene targeting is to exclude the expression of a gene. The transcribed RNA of the gene of interest can be targeted using gene silencing techniques like RNA interference (RNAi). The RNAi is the post transcriptional sequence specific targeting/ silencing of mRNA of the gene which inhibit the protein expression. The RNAi is activated upon the introduction of 19-21 nucleotide long double stranded (ds) RNA into the cells. On the silencing of mRNA of gene of interest by RNAi technology, the reduction of up to 90% in the protein expression of gene can be achieved. The gene can be targeted in the cells lines by the transfection of the siRNA or a vector expressing the shRNA in the cell lines. The gene is still present but the significant reduction of gene expression means the gene has almost lost its function. It is helpful in quick study of the function of gene of interest.

3. Selection of mouse as favorite models for gene knock out: Mouse has been considered as the most suitable and favorite animal model for generation of knock out animal for research because of relatively low cost of maintenance and generation time of nine weeks of the mouse, produce large number of offsprings, physiological and anatomical similarities between mice and humans, similarity of 97.5% in the mice and human working DNA, researcher ability to introduce the desired gene at germ line genome of the mouse. Besides that the molecular and stem cell biology have allowed researchers to create custom-made mice through gene targeting in mouse embryonic stem (ES) cells, human diseases such as cystic fibrosis and Alzheimer's. The diseases can be induced by manipulating the mouse genome and environment. The mice then naturally develop conditions that mimic human disease such as cardiovascular disease, cancer and diabetes etc. Nearly 11,000 genes have already been knocked out in mice and scientists are striving to knock out all ~25,000 mouse gene.

Gene Knockout in Mouse

Knockout mouse refers to the mouse in which one or more genes have been completely knocked out. The gene has to be completely inactivated from the whole body of the mouse. Therefore, the gene needs to be knocked out at embryonic stage in the embryonic stem (ES) cells.

The series of steps are to be followed for the generation of KO mouse:

1. Identification of gene of interest: The gene of interest need to be identified that needs to knock out. Once the gene selected for inactivation, the information of chromosomal locations, gene size and exon/ intron should be collected. The major restriction site to be identified and restriction

map of the genmic locus to be prepared to plan better cloning strategies in the replacement vector. The flanking region should also be examined to avoid any disruption of neighboring genes during the recombination process.

2. Construction of targeting vector: The efficient targeting replacement vector needs to be constructed to introduce the targeted mutations or to replace the gene of interest in ES cells. The drug resistance maker i.e. neomycin in the targeting replacement vector should be able to replace the essential coding region of the gene of interest in the genetic locus. The popular replacement vector possesses both positive and negative selection marker with two homology arms flank with positive drug selection marker (neor) for homologous recombination which is placed adjacent to the targeting arms and negative selection marker (HSV-tk) for screening the non-homologous recombination. The DNA strands of the vector and the cells are exchanged through the process of homologous recombination. At least 2 kb of sequence homology is needed on the arms of the targeting gene in the ES cell with arms of resistance gene in the vector for the recombination process to take place. Further, the homology of 6-14 kb considered as typical for the recombination. Therefore, the targeting vector should be designed by incorporating this important feature of homologous recombination to boost the chance of success while developing a KO animal. The homologous recombination is the key feature for generating the knockout mouse while non-homologous recombination for generation of transgenic animal.

3. Isolation of embryonic stem (ES) cells: Since the purpose is to produce an animal in which gene has been knocked out from the complete body therefore we need to target the ES cells which form the complete body. Embryonic stem (ES) cells are pluripotent cell lines having the capacity of self-renewal and differentiation. The ES cells are harvested from the fertilized embryo at the balstocyst stage 4-5 days post-fertilization. The inner mass of blastocyst has mass of 10-20 cells and that inner mass recovered and culture in suitable culture media for expansion. The ES cells recovered from the blastocyst of the fertilized egg of a female mouse and that mouse is called as donor mouse.

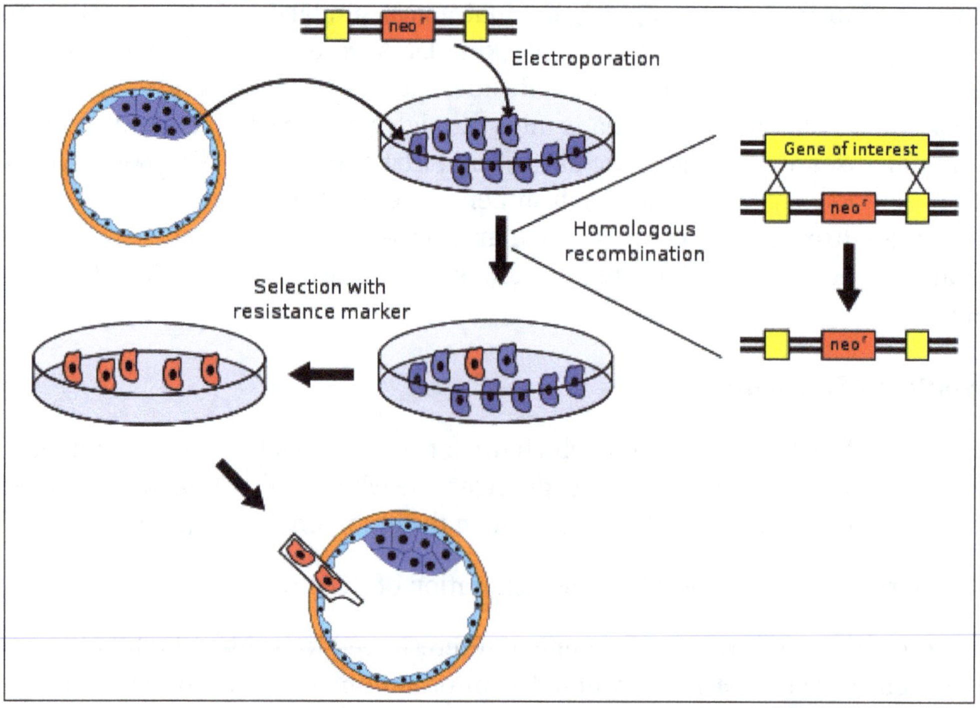

Figure: Steps to be followed for the generation of Knockout mouse by creating mixed genotype blastocyst.

4. Electroporation of vector into the ES cells: Electroporation is extremely efficient approach for transfecting the foreign DNA into the cells with the use of electrical pulse that generate temporary pores in the cell membrane. An electrical pulse for a few microseconds to a millisecond is passed through the cell suspension to disturb the phospholipid bilayer and form the temporary pores in the membrane. The raised electric potential across the cell membrane permit the charged DNA to pass through the pores of the membrane into the cells where homologous recombination take place.

5. Homologous Recombination in the cultured ES cells: Homologous recombination have important role in preserving and duplicating the genome by providing critical support for the replication of DNA and maintenance of telomere. The homologous recombination metabolizes DNA in all form of life and provides template dependent repair of DNA gaps, double stranded breaks and cross-linking of strand exchanged. Taking advantage of these properties of homologous recombination process, the scientist implies homologous recombination for exchanging of DNA strands between targeted replacement vector and gene of interest at designated genetic locus to inactivate the gene of interest. The double-strand break repair (DSBR)/ Holliday junction model and synthesis-dependent strand annealing (SDSA) are the two important models for homologous recombination.

6. Selection of ES cells with resistance marker: The targeted events by homologous recombination take place in very few cells or small percentage of cells and there is need to select the cells in which homologous recombination took place. This can be done through some resistance marker and the neomycin is the most common resistance marker. The resistance marker DNA strand of replacement vector get exchanged and incorporated at the designated genetic locus by replacing or knocking out the gene of interest and convert that particular cells resistance to that particular antibiotic. On culturing the ES cells in the presence of the antibiotic, the cells who does not possess the property of antibiotic resistance dies and only the cells having antibiotic resistance gene can survive.

7. Transfer of selected ES cells to the mouse: The selected ES cells transferred/injected to the host embryo of recipient/surrogate mouse for the development of pups. The pups developed are chimeras where gene of interest has been knocked out in some cells but some of the cells still having the gene. These chimeras need to be screen and confirmed before moving further. If the recipient embryo and the ES cells are from two different strains of mice with different skin will allows to visually selecting the chimeras just by color.

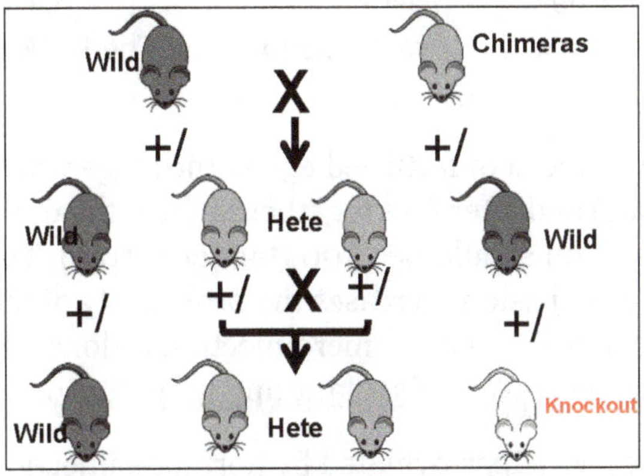

Figure: Mating of Chimera mouse with Wild for production of heterozygous and further mating of heterozygous to produce knockout mouse.

8. Mating to generate homozygous gene knockout mouse: The chimeric mouse mated with wild type mouse to incorporate the mutations in the germ line likely to produce 50 % heterozygous mouse. The heterozygous mouse carries one copy without gene (-) and one copy with (+) gene. The heterozygous mouse on mating with other heterozygous mouse ay produce 25% of the pups in which the gene of interest has been completely knocked out or loss the function of the gene.

Development of Transgenic Mouse/Animal

The transgene is introduction of foreign genes or altering the forms of an endogenous gene into an organism. Generally, transgene means is not replace of the endogenous gene, but rather the integration of additional copies of it. The introduced gene is called transgene and the organism as transgenic organism/animal. The transgenic animals are used to over-express the genes by introducing additional copies of DNA/gene to study its biological role. Transgenic animals randomly integrate the cloned genes/exogenous DNA in the host genome at non-homologous sites. The frequency of random integration is very high which help to produce the transgenic mice and the process is highly efficient and straight forward. The site of random integration of foreign DNA into the genome is not predetermined, therefore on injecting the same foreign DNA in the multiple embryos the transgene locus will be different in the each founder (mouse having transgene). However, transgenic animals which will be produced as next progeny of single founder will share the same transgene locus. The insertion of the transgene into a normally-functioning endogenous gene have unforeseen consequences by causing death, changes in physiology and phenotypic anomalies. Transgenic has a range of experimental and potential therapeutic applications.

Transgenic animals can be generated by manipulation in culture of ES cell by selecting the non-homologous recombinant in ES cells instead of selecting the homologous recombinant ES cells as discussed above. The non- homologous recombinant cells can be selected by culturing the cells under negative selection marker in the cell culture and incorporated back into the embryo at the blastocyst stage. The mouse developed can produce transgenic mouse on mating as discussed above in gene knockout.

Another technology to produce transgenic mouse is by microinjection of the foreign DNA into the pronuclear of the fertilized egg. We shall be discussing the other most popular technology for production of transgenic animal by microinjecting of the foreign DNA which is technically simpler than manipulating ES cells. The gene of interest that needs to be introduced or over-expressed can be cloned in some BAC or YAC vector for amplification but for pronuclear microinjection the fragment should be free from flanking plasmid sequences. The plasmid sequence may be toxic therefore DNA fragment must be without vector sequences. The DNA must be intact to avoid integration of any partial DNA.

The DNA is injected in the pronuclei of fertilized egg at the stage when ova has two nuclei, one of each gamete (male and female) and later fuse together to form the diploid nucleus. The number of copies of DNA for single insertion should be >100 (tandem array). The microinjection of DNA at 2 cells stage before the first cell division increases the probability of incorporation of transgene to 10-30 % of somatic and germ line cells and if microinjection is done after 2 cell stage the probability of incorporation of transgene to 10 % of somatic and germ line of developing chimera.

The egg donor female mouse gets super-ovulated by hormone injection to release more eggs than usually produce. In the 10 female mice, around 200 eggs can be produced. These female mice are mated with the male mice for fertilization and these fertilized eggs harvested next morning. Thereafter, the DNA is microinjected into the pronuclei of each of the egg. The pronuclei of mouse eggs can be visualized by Differential Interference Contrast (DIC) optics. These microinjected eggs are cultured overnight to the 2-cell stage and then transfer to the oviducts of pseudo-pregnant recipient's mouse (surrogate mother). The gestation period of mouse is 19-21 days.

Once pups are born, tissue samples (Tail) taken approximately after 7 days of birth and screened for the presence to tarnsgene. The mouse having transgene are called founder.

Development of Conditional Knockout Mouse

The constitutive KO of gene in mouse may sometime produce undesirable result or may be embryonic lethal. Therefore, there is need of conditional KO model where the conditional gene knockout enables the inactivation of a gene of interest in a specific tissue and/or at a specific time point while the other tissues of the mouse should retain its wild type function.

The Cre-loxP system is most commonly utilized for conditional mutation of a gene. The loxP are the inverted repeat of 13 bp DNA sequences separated by 8 bp of non-palindromic sequences. The size of loxP sequences is 34 bp. The DNA sequences of loxp site is 5'-ATA ACT TCG TAT A gc at ac at T ATA CGA AGT TAT-3', the sequences in capital letter are invertd repeat and in small letters/ underlined are spacer non pallindromic sequences. The exons that need to be excised in conditional manner should be flanked by two loxP sites. The cre recombinase is enzymes isolated from the P1 bacteriophage which mediated site specific recombination between two loxP sites.

The DNA sequences between two loxP site floxed by Cre-recombinase. Floxed refer to the excise of DNA flanked by two loxP sequences and fusion of DNA after excision by Cremediated recombination. The Cre mediated recombination depends upon the orientation of the loxP sites in the DNA. DNA sequence between two loxP sites oriented in the same direction excised as a circular loop of DNA while the DNA between two oppositely orientated loxP sites inverts the DNA. The Cre-recombinase can also catalyses the the process of translocation between the two different DNA strands having loxP site.

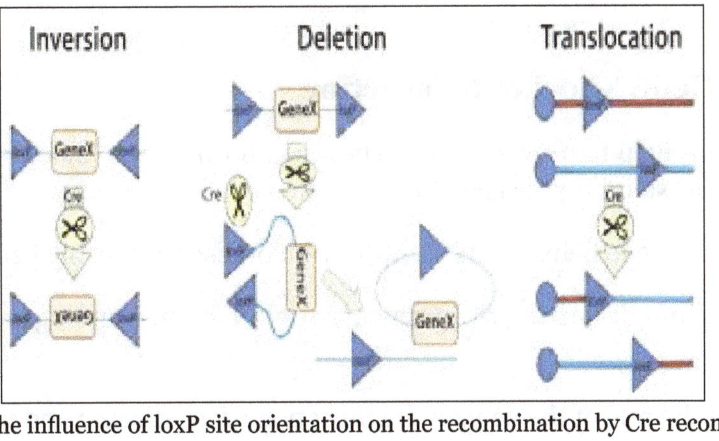

Figure: The influence of loxP site orientation on the recombination by Cre recombinase.

The loxP sites must be placed in the introns of the gene to be targeted so that it does not interfere with promoter, coding sequences and neighboring genes. These loxP mice (Floxed mice) contain loxP sites flanked exons in all tissues of the mice and these mice are phenotypically and physiologically wild before excision of exon by Cre recombinase. These floxed mice can be generated by the procedure as followed in gene knockout with slight modifications in vector design by incorporation of the loxP site. The Cre recombinase expressing transgenic mice can generated as described above in section of transgenic animals. The expression of Cre recombinase should be under the control of inducible promoter by which the promoter activity can be conditionally induced. Some of the examples are discussed here. Some examples of conditional knockout mouse model:

The recombination can be regulated by tamoxifen-inducible Cre expression system. The Cre expression is controlled by the mutated ligand binding domain of the estrogen receptor and this does not allow the transcription until the tamoxifen is present. Therefore, on applying the tamoxifen exogenously through the injection, the transcription and further expression of Cre-rembinase can be induced.

The recombination can be regulated by reverse tetracycline-controlled transactivator (rtTA) inducible Cre expression system. The rtTA can only induced the transcription in the presence of Doxycycline. Therefore, on applying the doxycycline exogenously through the injection, the transcription and further expression of Cre-rembinase can be induced. When and where the Cre needs to be expressed can controlled by the timing and site of injection of the inducible.

The SCL gene is expressed in the brain tissue which is critical for the embryonic hematopiosis and knock out of SCL at embryonic stage is lethal for the mouse. Therefore, the need of generation of the conditional model arises. The Cre expression was regulated under the control of Nestin promoter. Nestin is a pro-neural gene has specific expression in the brain tissue. Therefore, the Cre recombinase expressed only in brain and floxed the SCL while the gene is present in other tissue of the mouse body. This is an example of tissue specific conditional KO of the gene.

The conditional knockout mice have dramatically increased on the introduction of Cre/loxP gene targeting technology like Mx1 promotor-Cre, GR-Cre induced by interferon and dexamethasone respectively.

Applications of Gene Knockout/Targeting

- Gene knockout help to draw difference between normal and knockout organism which is helpful in understanding the gene function.
- Knocking out of gene helps in the identification of disease causing genes.
- Investigating and dissecting the pathways (Cellular, Immunological and Inflammatory etc.)
- Investigating the role of the gene in the causation of the disease and in the development of gene therapy.
- Transgenics helps in the development of biopharmaceuticals and Identification of the Cure of the disease.

Transgenic Livestock

Microinjection is the most common gene transfer method for transgenic livestock. Pronuclear DNA microinjection. DNA microinjection into pronuclei of zygotes is the classic method of gene transfer in farm animals and has remained basically unchanged since its development 15 years ago. This technique allows addition of DNA fragments which integrate randomly and is therefore not suitable for introducing targeted mutations in farm animal genomes. The principle of the procedure is similar to that used in mice, although species-specific modifications are necessary, particularly for the recovery of embryos, the microinjection process, and for the transfer of injected embryos to recipients. Whereas embryos from rabbits, pigs, sheep and goats are usually recovered from superovulated donor animals, in vitro maturation and fertilisation of oocytes is used to obtain a sufficient number of zygotes for gene transfer experiments in cattle. This makes the production of transgenic cattle economically feasible, although the overall efficiency is low. Due to their opaque cytoplasm, zygotes from cattle and pigs have to be centrifuged prior to injection to move the abundant cytoplasmic lipid granules to one side and make the pronuclei visible. The pronuclei of sheep embryos can be visualised by Nomarski phasecontrast optics.

Overall, the efficiency of DNA microinjection technique in farm animals is, as a rule, lower than in the mouse. Therefore, attempts have been made specifically in cattle to take biopsies from injected and cultured embryos and to prove the presence of injected sequences by polymerase chain reaction (PCR). Furthermore, expression-based approaches have been used to increase the proportion of transgenic embryos using either neomycin phosphotransferase (neo) or green fluorescent protein (GFP) expression as selectable markers.

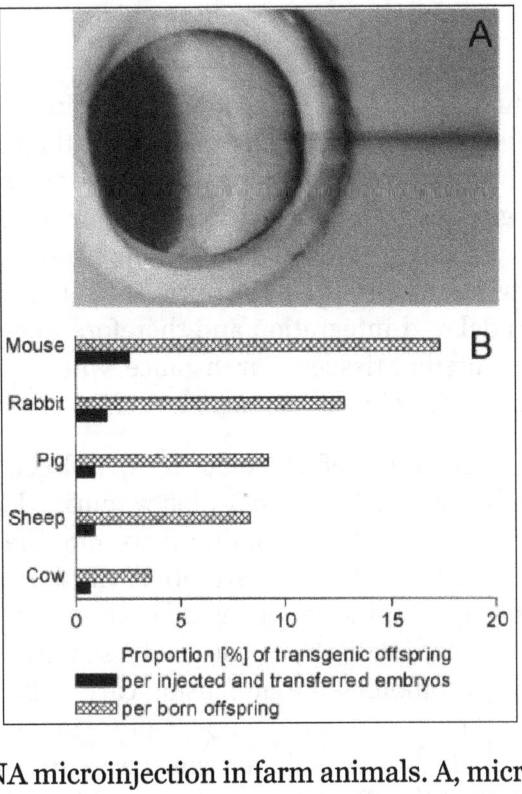

Figure shows pronuclear DNA microinjection in farm animals. A, microinjection into a porcine zygote in which cytoplasmic lipid granules have been moved to one side by centrifugation to visualise pronuclei. B, efficiency of the DNA microinjection technique in various species.

Random integration of microinjected DNA bears the risk of causing insertional mutations and varying expression due to effects of the adjacent DNA sequences. These may cause aberrant expression patterns, reduced or complete lack of expression. One possibility for overcoming the latter problem is the transfer of large DNA fragments, forming functionally independent chromatin domains. The first successful example for this strategy in livestock was the generation of transgenic rabbits which harboured yeast artificial chromosomes (YACs) coding for tyrosinase, thus rescuing albinism. The potential of YAC-based vectors for high level expression of therapeutic proteins in the mammary gland has recently been demonstrated in rats.

Another approach for overcoming position effects is the use of matrix attachment regions (MARs) or scaffold attachment regions (SARs) which can bind the nuclear matrix and/or chromosomal scaffold and act as boundary elements (insulators), protecting the transferred DNA constructs from regulatory effects of the neighbouring chromatin, thereby enabling copy number-dependent, position-independent expression of transgenes. Although very attractive in theory, the effects of MARs were not in all studies as expected.

Following DNA microinjection, embryos are transferred to synchronised recipients, which may be done immediately or after culture in vitro. Embryo transfer has been facilitated by the development of elegant minimally invasive techniques, reducing stress to the recipient animals and maximising pregnancy and embryo survival rates. Table shows gestation time, age of sexual maturity, and period up to F1 animals which can be characterised for expression levels and patterns for several farm animal species. In spite of the overall low efficiency, most of the existing transgenic farm animals have been produced by the DNA microinjection technique.

Viral Vectors

Retroviral infection was the first method used to produce transgenic mice. Using retroviral vectors, genetic information is transferred as an RNA molecule which is then reverse transcribed and integrated into the host genome by means of the retroviral integrase and specific nucleotide sequences at the ends of the retroviral genome. Most retroviruses can only integrate in dividing cells, since this process depends on the breakdown of the nuclear envelope during mitosis, allowing translocation of the retroviral pre-integration complex to the nucleus. Thus, infection of early embryos with retroviral vectors resulted in delayed integration and therefore in genetic mosaics characterised by different insertion sites in different tissues. For instance, when retroviral packaging cells were placed under the zona pellucida of bovine embryos, mosaic fetuses were obtained.

To avoid mosaicism and to increase the efficiency of retroviral gene transfer in cattle, Chan et al. injected replication-defective vectors based on Moloney murine leukaemia virus, pseudotyped with the envelope glycoprotein of vesicular stomatitis virus, into the perivitelline space of either metaphase II (MII) oocytes (16 h after onset of maturation) or of zygotes. The rationale for using MII oocytes was that in this stage the nuclear envelope is absent and the oocyte remains in this state for much longer time compared with the M-phase of a somatic cell. Staining for the activity of a Rous sarcoma virus (RSV) promoter-neo/lacZ fusion carried by the vectors revealed 56 and 22% positive embryos produced by in vitro fertilisation/culture in the oocyte and zygote treatment groups, respectively. For the production of embryos which were transferred to recipients, another reporter gene coding for the Hepatitis B surface antigen was included in the retroviral vectors, which were again injected into the perivitelline space of MII oocytes or of zygotes. All four born calves from the former and 1 of 2 born calves and 2 late-stage aborted fetuses from the latter group were shown to be transgenic. Investigation for germ-line transmission in male calves was positive, whereas expression analyses showed either low or no activity of the reporter gene. Therefore, a remarkable efficiency of gene transfer was achieved by retroviral infection of MII oocytes, whereas proper expression of transgenes transferred in this way has still to be demonstrated. Major drawbacks of using retroviral vectors are the limited size (< 10 kb) of sequence that can be transferred because of physical limitations in volume of virus particles. This prohibits the use of larger regulatory elements or genomic constructs which are usually expressed at higher levels than cDNA constructs. Furthermore, the long terminal repeats (LTRs) flanking retroviral genomes have been reported to interfere with mammalian promoters, suppressing or misdirecting expression.

Table: Age of sexual maturity, gestation length and period up to the birth of presumed transgenic F1 generation in different livestock species.

Species	Sexual maturity (months)	Gestation length (days)	Generation interval (birth of F1 animals) (months)
Rabbit	3–4	30–32	6
Pig	5–8	114	12
Sheep	3–6 (m) 5–10 (f)	150	18
Goat	5–9 (m) 8–10 (f)	150	19
Cattle	811	280	30
Abbreviations: m, male; f, female.			

In addition to retroviral vectors, replication-defective adenoviruses have successfully been used to infect zona-free zygotes and to generate transgenic mice. These experiments revealed stable single-copy integration, transmission to offspring and expression of the carried lacZ reporter gene. A subsequent study investigated effects of timing of infection and titre of a replication-defective human adenovirus carrying a CMV-lacZ cassette on expression of the reporter gene following injection into the perivitelline space of mouse, rat or cow embryos. This study showed that depending on virus concentration a high proportion of treated embryos (up to 96, 76 and 58% of mouse, rat, and bovine embryos, respectively) expressed the reporter gene, with decreasing efficiency when the injection was performed into more advanced stages of embryos. Stable integration of adenoviral vectors into the genome of infected embryos remains to be proven for other mammalian species than mouse.

Sperm-mediated Gene Transfer

The idea of using sperm cells as vectors for introducing exogenous DNA into oocytes was spectacular, but the initial expectations generated by a publication of Lavitrano et al. have not thus far been fulfilled. Brackett et al. were the first to demonstrate that mammalian spermatozoa can bind exogenous DNA molecules. Since then, numerous papers reported binding of exogenous DNA to spermatozoa of amphibians, molluscs, fish, insects, birds, and mice as well as other mammals. A model has been proposed whereby DNA binding to sperm is conferred by specific DNA binding proteins (DBPs) present on the sperm surface (in the post-acrosomal region), however, the interaction between DNA and DBPs can be blocked by an inhibiting component (IF-1) in the seminal plasma. CD4 and major histocompatibility complex class II (MHC II) play distinct roles in the process of sperm-exogenous DNA interaction. Studies using knockout mice revealed that MHC II expression is required during spermatogenesis to produce sperm cells capable of taking up exogenous DNA, while CD4 is necessary for nuclear internalisation. Studies in mouse suggested, that plasmid DNA internalised in ejaculated spermatozoa associates with the nuclear scaffold, is extensively rearranged, and undergoes recombination with genomic DNA. In contrast, internalisation of DNA by epididymal sperm induced a metabolically active process similar to apoptosis.

In mammals, the results of sperm-mediated gene transfer are controversial due to the low efficiency and repeatability. In a large-scale experiment, Maione et al. obtained 7.4% transgenic offspring on average using mouse spermatozoa which were incubated with plasmid DNA during capacitation and subsequently co-incubated with oocytes. However, transgenic offspring were only obtained in 13 of 75 trials, some with an efficiency of more than 85%. The circumstances leading to these highly variable results remain to be determined. Artificial insemination in cattle with DNA-treated sperm resulted in a calf exhibiting weak transgene-specific signals of unexpected length, suggesting that less than one copy of a truncated construct per cell was present. Recently, generation of human decay-accelerating factor (hDAF)transgenic pigs by sperm-mediated gene transfer has been reported.

A modified and efficient technique of sperm-mediated gene transfer in mice was recently described. Intact or membrane-disrupted spermatozoa were shown to bind exogenous DNA (GFP or lacZ reporter constructs) and were used for intracytoplasmic sperm injection (ICSI) into unfertilised oocytes, resulting in a high proportion (6494%) of embryos expressing exogenous DNA. After embryo transfer, 1721% of recovered offspring expressed the integrated exogenous DNA. Moreover, 8 of 11 tested founders transmitted the transgenes to progeny. Although promising in mice, the application of this technique for the production of transgenic livestock is hampered by the overall low efficiency of ICSI in these species. Only recently, a reasonable proportion of pig embryos expressing a reporter construct have been obtained following ICSI of sperm coin-cubated with DNA. However, development of these embryos in vivo and stable transgene integration/expression remains to be shown.

Another possibility for producing transgenic animals in a sperm-mediated manner is the intratesticular transfection of germ cells. Mice and pigs were treated with busulfan, an alkylating agent, to destroy the developing male germ cells, and liposome/bacterial lacZ gene complexes were introduced into each seminiferous tubule using a microinjection needle. In mice, 8.014.8% of the seminiferous tubules expressed the introduced lacZ gene, and 713% of epididymal spermatozoa were confirmed by PCR as having foreign DNA. In pigs, exogenous DNA was also incorporated efficiently into male germ cells, and 1525% of the seminiferous tubules containing germ cells expressed the lacZ gene. However, so far none of the described sperm-mediated gene transfer techniques can be regarded as routinely applicable to mammalian livestock species.

Pluripotent Stem Cells

Pluripotent stem cells are not able to form an entire organism on their own, but have the capacity to develop into many cell types (including germ-line cells) following introduction into preimplantation embryos by aggregation with morulae or injection into blastocysts.

The first cell lines that fulfilled these criteria were isolated from teratocarcinomas of mice and were designated EC (embryonic carcinoma) cells. However, their use for transgenic technology is hampered by their unstable karyotype and consequently their low rate of germ-line transmission.

Embryonic stem (ES) cells, another type of pluripotent cells in mouse, are in most studies derived from the inner cell mass of blastocysts, although generation of ES cells from blastomeres of morulae or from eight-cell embryos has been reported. In contrast to EC cells, the ability of mouse ES cells to colonise the germ line is high, especially if the cells have been obtained from specific strains, such as 129 or C57BL/6. Furthermore, selection strategies against differentiated cells have been established, allowing the derivation of ES cells also from normally non-permissive strains.

Mouse ES cells have become one of the most important tools in experimental genetics, since targeted alterations of the genome can be induced in cultured cells by homologous recombination. This provides the technical basis for general or tissue-specific/inducible gene knockouts, site-directed insertions into defined loci, or the possibility of engineering large chromosomal rearrangements.

A third type of pluripotent cells from mice, embryonic germ (EG) cells, is derived from primordial germ (PG) cells and differentiates in vitro and in vivo in a similar way to ES cells. However, EG cell lines have been less widely used than ES cells to manipulate the mouse genome.

In view of the wide spectrum of applications for pluripotent cell lines in mouse genetics, numerous attempts have been made to establish equivalent cells from livestock species. Many different culture systems using various types of feeder cells and species-homologous or heterologous differentiationinhibiting cytokines have been tested, and ESlike'cells were reported for a number of species, including sheep, pig, cattle, rabbit, and goat. However, the characterisation of the cells was mainly based on morphological criteria and on expression of differentiation markers. Formation of chimeric animals was reported in pig and rabbit; however germline chimerism could not be demonstrated. In a recent experiment, ESlike cells were derived from a transgenic nuclear transfer embryo which had been produced by using a transfected fibroblast as nuclear donor. The ESlike cells were injected into morulae and transferred to recipients. Six of seven born calves were found to be chimeras, demonstrating that the transgenic ESlike cells had the potential to contribute to tissues from all three germ layers. In another experiment evaluating the in vivo pluripotency of bovine ESlike cells, they were aggregated with tetraploid eight-cell embryos and the aggregates transferred to recipients. Two of six resulting calves were shown to be chimeras, both of which died prohibiting evaluation of germline chimerism.

Other attempts concentrate on the use of PG cells as starting material for the generation of pluripotent cell lines from livestock species. The major advantage of this approach is the higher number of PG cells that can be recovered from a fetus compared with the number of undifferentiated cells available in a preimplantation embryo. We have shown that for initial passages of porcine PG cells, feeder layers expressing porcine stem cell factor (SCF) are advantageous. PG cell derived cell lines were established from whey acidic protein (WAP) human growth hormone (hGH) transgenic pig fetuses and contributed to chimeric animals produced by blastocyst injection produced transgenic porcine chimeras using transfected PG cell colonies. However, germ-line transmission could not be shown in either study.

It is still unclear why germ-line competent pluripotent cell lines could not be established from any other mammalian species than mouse so far. Problems with ungulate embryos include the prolonged preimplantation time, a rapidly proliferating trophoblast, and an early onset of differentiation in the mitotically quiescent ICM. Recently, less effort has been put into the establishment of germ-line competent cell lines from livestock species, since nuclear transfer using transfected donor cells provides an attractive alternative for targeted gene transfer.

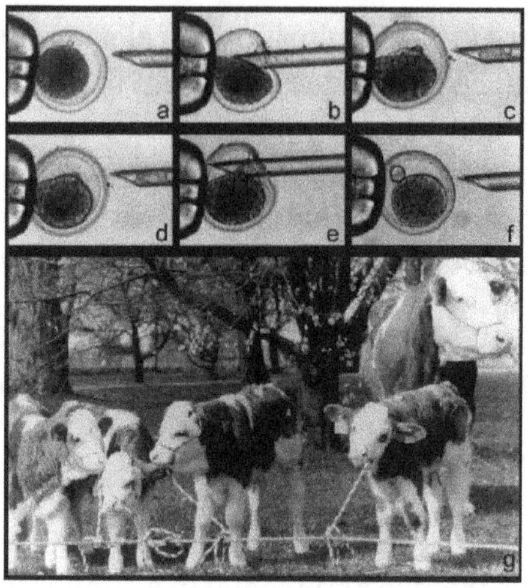

Figure shows nuclear transfer in cattle. a-c, enucleation of a recipient oocyte to produce a cytoplast. d-f, transfer of a nuclear donor cell (karyoplast) under the zona pellucida of the cytoplast. g, clone of four Simmental calves produced following electrofusion of karyoplastcytoplast complexes, in vitro culture of reconstructed embryos and transfer to recipients.

Nuclear Transfer using Transfected Cells

Nuclear transfer (NT) technology involves the transfer of a donor nucleus (karyoplast) into the cytoplasm of a zygote or a metaphase II oocyte from which the genetic material has been removed. If nuclear transfer is done by electrofusion of karyoplast and cytoplast, both the nucleus and cytoplasmic components of the donor cell, including the mitochondria, are transferred. The latter may persist in the reconstructed embryos, developing fetuses and born animals, although heteroplasmy has not been detected in cloned sheep.

Initial nuclear transfer experiments in livestock used blastomeres from pre-implantation embryos as nuclear donors, limiting practical applications in biotechnology. However, pioneering experiments demonstrated first that ovine embryonic cells can be expanded in culture before using them for nuclear transfer, and later that not only cells of embryonic origin, but also cells derived from fetal or adult tissues can be successfully used for this technology. In the meantime, nuclear transfer using differentiated donor cells has been achieved in sheep, cattle, goat, pig, and mouse. Table summarises efficiency data for nuclear transfer in cattle using different types of donor cells.

Table: Efficiency of nuclear transfer in cattle using different types of donor cells.

Cell type	Fused n (%)	Cleaved n (%)	Blastocysts n (%)	Transferred blastocysts n	Pregnancies nP/nT (%)	Calves n (%)*
PFF	496 (76)	273 (72)	115 (30)	23	8/12 (67)	2 (9)
PGC	507 (84)	283 (66)	115 (27)	32	8/17 (47)	1 (3)
MGCL	118 (57)	31 (46)	0 (0)	–	–	–

MGPC	223 (63)	93 (66)	36 (26)	4	2/2 (100)	1 (25)
PAF	92 (89)	70 (85)	49 (60)	16	5/12 (42)	1 (6)
PGrC	109 (85)	64 (69)	20 (22)	6	1/3 (42)	2 (33)
PMC	112 (71)	49 (62)	10 (17)	4	0/2 (0)	–
TFF	511 (85)	328 (75)	144 (33)	95	24/49 (49)	8 (8)
nTFF	177 (94)	132 (79)	87 (52)	44	11/21 (52)	13 (30)

*Based on the number of transferred blastocysts. Abbreviations: PFF, primary fetal fibroblasts; PGC, primordial germ cells; MGCL, mammary gland cell line (immortal); MGPC, mammary gland primary cells; PAF, primary adult ear fibroblasts; PGrC, primary granulosa cells; PMC, primary milk cells; TFF, transgenic fetal fibroblasts; nTFF nontransgenic fetal fibroblasts.

The implications for biotechnology of nuclear transfer using cultured cells are twofold: (i) the pool of genetically identical donor cells is not anymore limited to the number of cells in an early embryo; and (ii) cultured cells can be transfected before using them for nuclear transfer, providing a new route for the generation of transgenic animals.

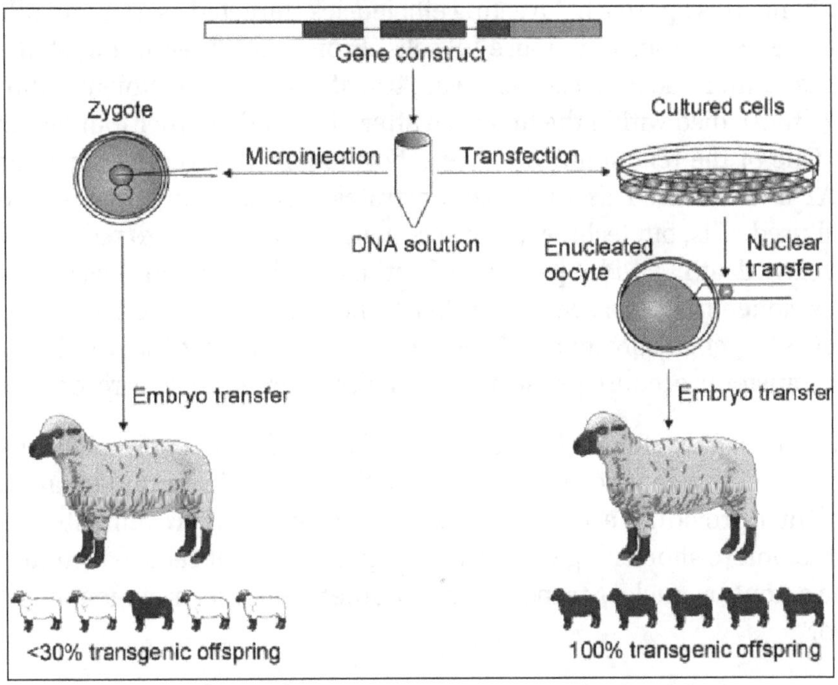

Figure: Genetic modification of farm animals by DNA microinjection vs. nuclear transfer using transfected donor cells.

The proof of principle for this strategy was the generation of human factor IX transgenic sheep by nuclear transfer using transfected fetal fibroblasts. In this study, 20.8 sheep where required on average to produce one founder animal, which was less than half the animals needed when using the DNA microinjection technique under comparable conditions (51.4 sheep per founder animal). Shortly afterwards the first cloned transgenic calves were reported, and more recently transgenic goats were produced by nuclear transfer using cells derived from a transgenic fetus obtained after artificial insemination of a non-transgenic adult female with semen from a transgenic male.

Transgenesis by nuclear transfer using transfected cells provides a number of advantages over the classic DNA microinjection technique: (i) male or female cell lines can be used, depending on the aim of a particular project; (ii) integration and eventually expression of the transferred sequences can be investigated in cultured cells before using them as nuclear donors; (iii) transfected cells can be stored frozen; (iv) all cells within a cloned animal contain the introduced gene thus eliminating mosaicism and ensuring germ-line transmission; and (v) the time, costs, and number of animals involved are reduced. A production flock of transgenic animals can be generated in a shorter time since one generation of breeding is eliminated, and multiple genomic copies can be produced from a stable transfected cell line.

Most importantly, transgenesis by nuclear transfer offers a route for targeted genetic modifications in livestock species, eventually providing the methodological pendent to ES cell technology in mice. Recently, a mammary gland-specific expression vector for á1antitrypsin (AAT) was targeted into the α1(I) procollagen (COL1AI) locus of ovine fetal fibroblasts which were subsequently used for nuclear transfer to produce the first gene-targeted sheep. The target locus (COL1AI) was chosen since this gene is highly expressed in fetal fibroblasts, allowing the use of a promoter-trap strategy, i.e. a promoter-less neo selection marker which was only expressed when integrated downstream of an active promoter. Using this approach, targeting efficiencies (targeted vs. random integration events) were similar to those usually seen in mouse ES cells. From four different targeted cell lines, a total of 80 nuclear transfer morulae or blastocysts were transferred to 42 recipients who gave birth to 14 live lambs. Half of those died within the first 30 h after birth and another four lambs within the first 12 weeks of life. One of the remaining lambs was hormonally induced to lactate and showed high expression of AAT in the milk. This study demonstrates that gene targeting is possible by nuclear transfer using cultured cells, but technical improvements are necessary to reduce the high incidence of abortions and early death. A wide spectrum of pathological alterations has also been observed in fetuses and calves cloned from transgenic fibroblasts, however abnormalities are not limited to nuclear transfer clones of genetically manipulated cells, suggesting that the problems to be overcome lie in the nuclear transfer procedure per se or in the culture conditions for reconstructed embryos.

Quantitative analysis, e.g. by RT-real time-PCR analysis, of gene expression profiles in nuclear transfer embryos vs. embryos recovered by fertilisation will help to understand mechanisms of reprogramming, thus providing a rational basis for technical improvements. Furthermore, the consequences of telomere shortening during prolonged culture of cells, which may or may not be reversed to some extent by nuclear transfer, and of other genomic or epigenomic changes remain to be determined.

Transgenic Fish

A transgenic fish is one that contains genes from another species. A transgenic fish is an improved variety of fish provided with one or more desirable foreign gene for the purpose of enhancing fish quality, growth, resistance and productivity. Typically, genes of one or more donor-species are isolated, and spliced into artificially constructed infectious agents, which act as vectors to carry the genes into the cells of recipient species. Once inside a cell, the vector carrying the genes will insert into the cell's genome.

A transgenic organism is regenerated from each transformed cell (or egg, in the case of animals), which has taken up the foreign genes. And from that organism, a transgenic variety can be bred. In this way, genes can be transferred between distant species, which would never interbreed in nature.

The application of genetic engineering to animals, like potatoes with built-in insecticide, could provide numerous benefits, including the possibility of a safer, cheaper food supply and the creation of new sources for inadequate pharmaceutical resources. With the advancement in the field of genetic engineering, the application of its commercial use has also increased. Aquatic animals are being engineered to increase aquaculture production.

The use of genetic engineering and rDNA technology has done miracles in medical and industrial research. The transgenic fish are being promoted as the first marketable transgenic animals for human consumption. One of the most important aspects between fish and other terrestrial animals for cultivation and genetic improvement is that, usually, fishes have higher levels of genetic variation and hence more scopes for selection than most mammals or birds.

Using the gene transfer technology, scientists now have created a genetically engineered variety of Atlantic salmon that grow to market size in about 18 months, otherwise the fish takes about 24-30 months for becoming market size fish. It is also hoped that we can now modify a large number of fishes with fast growing characteristics and bring Blue Revolution. The following are the important points needed for genetic engineering (gene transfer) to produce transgenic fish:

- A gene sequence is to isolate for the particular characteristics; for example, growth hormone gene.
- These genes (gene sequence) are then inserted into a circular DNA known as plasmid Vector (enzymes endonucleases and ligases are used).
- Plasmids are harvested in the bacteria to produced billions of copies.
- Plasmids are introduced into linear DNA. The linear DNA is sometimes called a gene cassette because it contains several sets of genetic material in addition to new inserted gene; for example, growth hormone gene. The technology is available to integrate genes in germ line of developing individual (fish) and finally transmitted into further generations.
- Making the cassette a permanent part of fish's genetic makeup.

Development of Transgenic Fishes

Development of transgenic fish has focused on a few species including salmon, trout, carp, tilapia and a few others. Salmon and trout are cash crops while the others primarily provide sources of protein. Currently, about 40 or 50 labs around the world are working on the development of transgenic fish.

About a dozen of them are in the U.S., another dozen in China, and the rest in Canada, Australia, New Zealand, Israel, Brazil, Cuba, Japan, Singapore, Malaysia, and several other countries. Some of these labs are associated with companies that expect to commercialize their fish in a few more years. Many of the fish under development are being modified to grow faster than their wild or traditionally bred aquaculture siblings.

Faster growth is usually accomplished by transferring a fish growth hormone gene from one species of fish into another. The faster growing fish not only reach market size in a shorter time, they also feed more efficiently. Trout growth hormone (GH) was used to produce transgenic carp with improved dressing properties. Such transgenic carp are recommended for production in earthen ponds.

Transgenic Salmon

The Atlantic salmon is engineered with a pacific salmon, growth hormone driven by the arctic antifreeze promoter gene. The rapid growth of that transgenic salmon is achieved, not so much by the transgenic growth hormone as by the antifreeze gene promoter that functions in the cool water desirable for salmon flavour.

The transgenic Coho grew on average 11 times faster than unmodified fish and the largest fish grew 37 times faster. The growth hormone levels in the transgenic fish are high year-round, rather than falling off in the winter as occurs in ordinary salmon. The modified salmon are large enough to be marketed after one year, in contrast to standard farmed salmon that do not reach market size for at least three years.

Transgenic Tilapia

Tilapia fish, native to Africa, are cultured world-wide as "poor man's food", second only to carp as warm water food fish, and exceeding the production of Atlantic salmon (whose market value is twice that of tilapia). Tilapia has been extensively genetically modified and promoted as a transgenic fish exclusive for isolated or contained production.

Transgenic tilapia, which is modified with pig growth-hormone, has three times larger than their non-transgenic siblings. Tilapia genetically modified with human insulin grew faster than non-transgenic siblings, and could also serve as a source of islet cells for transplantation to human subjects.

Transgenic Medaka Fish

Purdue animal scientist Muir and Howard used tiny Japanese fish, Oryzias latipes called medaka to examine what would happen if male med akas genetically modified with growth hormone from Atlantic salmon. Inserting a gene construct consisting of the human growth hormone driven by the salmon growth promoter into medaka produced the transgenic medaka.

The viability of groups of modified and conventional fish was measured at three days of age, and 30 percent fewer transgenic fish survived to that age. The researchers calculated that large males had a four-fold mating advantage, based on observations of wild-type medaka. In another experiment Silk moth genes were introduced into Medaka fish to create resistance to bacterial pathogens.

Transgenic Zebra Fish

The tiny zebra fish (Bmchydanio rerio) that lives in aquariums, was genetically modified to produce a fluorescent red pigment, and is being promoted for sale as a household aquarium pet, the "goldfish". The goldfish caused a stir in the United States because regulation of such transgenic pets is murky and none of the major regulatory agencies: Food and Drug Administration (FDA),

United States Department of Agriculture (USDA) or Environment Protection Agency (EPA), has been willing to take the lead in regulating the goldfish (even though USDA does deal with pet animals). The goldfish is available for sale from January 5, 2004 without regulatory approval in United States.

Figure: Zebra fish Glofish genetically modified to produced fluorescent red pigment.

Gong developed novel varieties of the Zebra fish. Three “living colour” fluorescent proteins, green fluorescent protein (GFP), Yellow fluorescent protein (YFP), and red fluorescent protein (RFP or dsRed), were expressed under a strong muscle-specific mylz2 promoter in stable lines of transgenic zebra fish.

These transgenic zebra fish with vivid fluorescent colours (green, yellow, red or orange) fluorescent proteins can be seen with naked eyes under both daylight and ultraviolet light in dark. The green fluorescent protein (GFP) is originally isolated from the jellyfish (Aequorea tictoria).

Auto-transgenesis

Indian scientists are concentrating on developing transgenic fish through auto-transgenesis which involves just increasing the copies of growth hormone genes present in a fish as opposed to allotransgenesis which amounts to transfer of genes from different species.

The increase in growth homone genes leads to an increase in flesh content. Indian scientists feel that auto-transgenesis is safer and less controversial. According to T.J. Pandian of the school of biological sciences in the Madurai Kamaraj University, the generation time of most fish species is shorter and breeding frequency is relatively higher.

A single female can produce several hundred or thousand eggs and thus provide a larger number of genetically identical eggs. Besides, the most important advantage is that the fertilization is external and can be readily controlled by experimental manipulation. According to Pandian, “the limited availability of transgenes of piscine origin had been the major hurdle in production of transgenic fish. However, with advancements in molecular biology, more than 8500 genes and cDNA sequences of piscine origin have been isolated, characterized and cloned in the world.”

Controlled Culture of Transgenic Fish and Feed

Pond commercial culture is effective for carp and tilapia, but more difficult with salmon and trout. Currently, pond culture is suitable for carp and tilapia because the fish are vegetarians, carnivorous salmon and trout depend on a diet of fish and fishmeal but the worldwide stock of feed fish has diminished and suitable vegetable meat substitutes must be found. Atlantic salmon (as typical cold water carnivores) cannot thrive on a diet of rapeseed oils but the fish can achieve maturity if finished with fish oils at least 20 weeks near the end of their maturity cycle. GM oil rapeseed with enhanced production of long chain fatty acids are proposed to serve as feed for pond cultured fish. And glyphosate- tolerant GM canola meal has been pronounced substantially equivalent to non-GM canola as feed for rainbow trout.

Gene Transfer Technology for Development of Transgenic Fishes

The most commonly used methods in fish biotechnology are chromosome manipulation and hormone treatments, which can be produced triploid, tetraploid, haploid, gynogenetic and androgenetic fish. Other popular methods of gene transfer in fish are microinjection, electroporation of sperms, electroporation of eggs and incubation of sperms. Following are the main steps in gene transfer for development of transgenic fish.

Preparation of DNA Construct

Desired transgene should be a recombinant gene or DNA construct, which is constructed in plasmid that contains an appropriate promoter-enhancer element and a structural DNA sequence. The foreign genes are typically introduced with strong genetic signals, promoters and/or enhancers, which enable the foreign genes to be expressed at very high levels continuously (or constitutively), effectively placing those genes outside the normal metabolic regulation of the cell, and of the transgenic organism resulting from the transformed cell.

Three Main Types of Transgenes

- Gain-of-Function: These transgenes are able to increase particular function in transgenic individual after their expression. For example growth hormone genes from mammal and fish linked to appropriate promoter-enhancer element and a structural DNA sequence to produce GH transgene. This GH transgene when express in transgenic individuals increases production of growth hormone leading to enhanced growth of transgenic animal.
- Reporter Function: These transgenes are able to identify and measure the strength of promoter-enhancer element.
- Loss of Function: This transgene is not yet used for modification of transgenic fish. Such transgenes are used for interfering with the expression of host genes. The promoter-enhancer elements of transgenes are linked to a growth hormone gene of fish.

Hence transgenic fish contain extra DNA sequences that are originally derived from same species. Gene construct is then introduced into fertilized egg or embryo, so that transgene be linked to genome of each cell of egg or embryo.

Gene Transfer by Microinjection

Microinjection is most successfully and widely used technique for gene transfer in fish. One method of microinjection technique involves the use of fine injection needle for introducing DNA into cut site in the cell. While doing so it destroys those cells that are in direct contact with the injected DNA.

To ensure the integration of the DNA it should be injected to intact cells close to the cut site. The injection apparatus consists of a dissecting stereomicroscope and two micromanipulators, one with a glass micro-needle for delivering transgene and other with a micropipette for holding fish embryo in place.

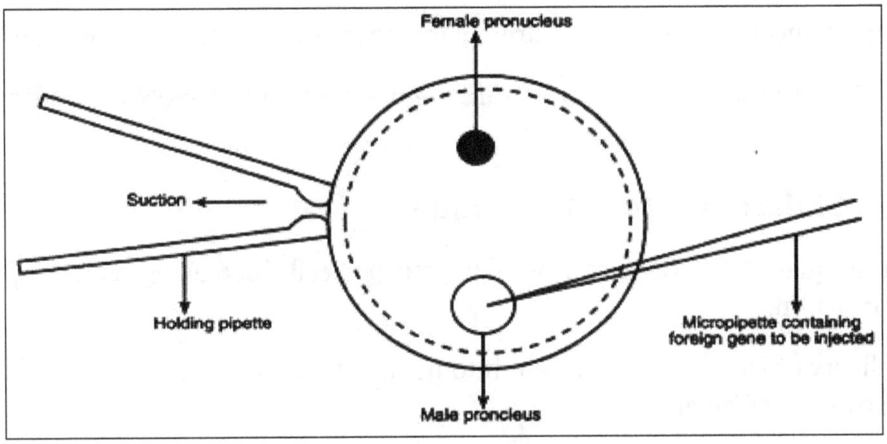

Figure: Micro injection apparatus to introduce DNA in intact Cell.

The success of microinjection technique depends on the nature of egg chorion. The soft chorion facilitates the microinjection while the thick chorion limits the ability to visualize the target for injection of DNA. In many fishes (Atlantic salmon and rainbow trout) the egg chorion gets tough and hard just after the fertilization or to contact with the water and provides a difficulty in injecting the DNA.

But using the following methods can solve this problem:

- By using the micropyle (an opening on the egg surface for sperm entry during the fertilization) for inserting the injecting needle.
- By using microsurgery for making an opening on the chorion.
- By digesting the chorion with enzymes.
- By using 1mM glutathione for initiating fertilization and reducing hardness of chorion.
- By direct injection to the unfertilized eggs.

Another technique of gene transfer is intra-nuclear microinjection, which involves direct physical approach using a fine needle to deliver DNA into cell or even nuclei. To facilitate rate of microinjection protoplast with partially reformed cell wall may be attached to a solid support with artificially bound substrate - without damaging the cells. Solid support may be of either glass cover slips or slides.

Steps of Microinjection Technique:

- Desired eggs and sperms are stored separately at the optimum conditions.
- Add water and sperms and initiate the fertilization.
- Ten minutes after the fertilization, eggs are dechorionated by trypsinization.
- Fertilized eggs are microinjected with desired DNA just within a few hours of fertilization. DNA is released into the centre of the germinal disc to the first cleavage in dechorionated eggs. The time available for microinjection is first 25 minutes and that too between fertilization and first cleavage.
- After microinjection the embryos are incubated in water until hatching takes place.

Survival rates of microinjected fish embryos are seem to be about 30-80% depending the fish species.

Advantages of Microinjection Technique

- Optimum quantity of DNA can be delivered per cell, increasing chances for integrative transformation.
- The delivery of DNA is precise, even into nuclei of target cell again improving chances for integrative transformation.
- The small structure can be injected.
- It is a direct physical approach; hence it is a host range independent.

Disadvantages of Microinjection Technique

- A single cell can be injected at a time; hence it is time consuming process.
- It requires sophisticated instruments and specialized skills.
- Limited embryonic time restricts injection to more eggs and a low transformation rate.

Gene Transfer by Electroporation

It is a simple, fast, efficient and convenient method for transferring gene. This method involves an electrical pulse to deliver DNA into cells. The cells are exposed to a short electrical shock, which make the cell membrane temporarily permeable to DNA.

The desired DNA fragment is placed in direct contact of protoplast membrane, which enters into the cell upon electric shock. Hole may be created as a result and stabilized by a favourable dipole interaction with electric field.

Electroporation involves a chain of electrical pulses for permeation of cell membrane, thereby allowing the entry of DNA into fertilized eggs. The rate of DNA integration in electroporated embryo is more than 25% is the surviving rate, which is slightly higher in comparison of microinjected ones.

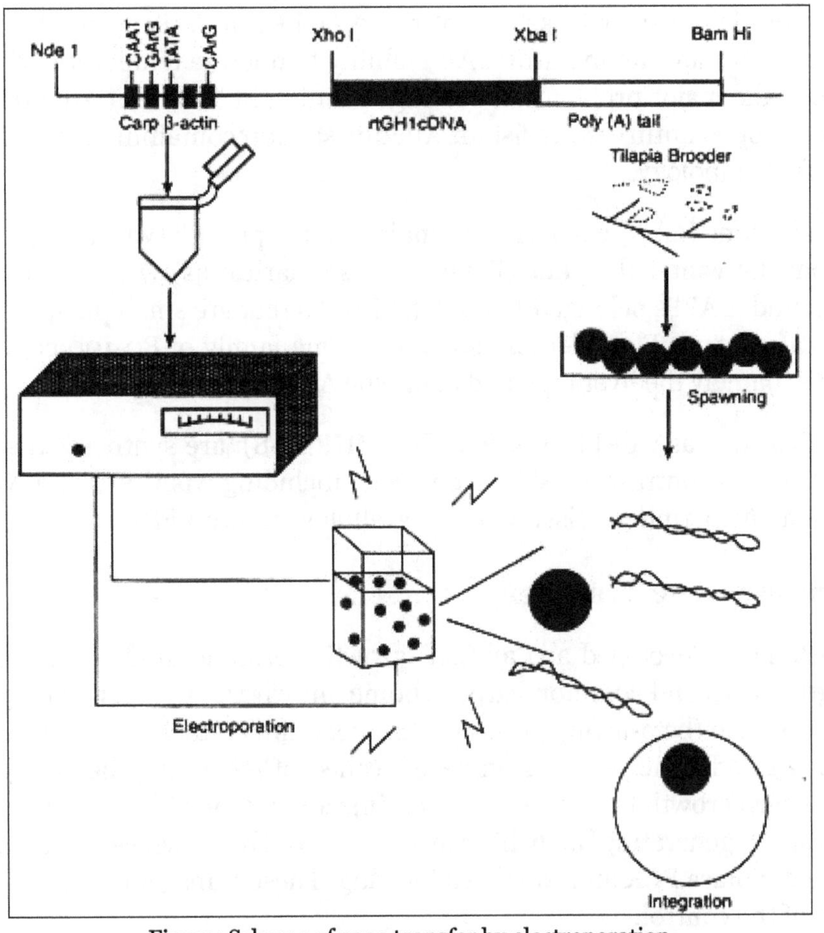

Figure: Scheme of gene transfer by electroporation.

Advantages of Electroporation Technique

- It allows simultaneous entry of DNA constructs.
- It is more suitable method for those species, which has very small eggs for microinjection.
- This method does not require specialized skill.

Antifreeze Protein Gene Transfer

Many teleost inhabiting icy marine water in the Polar Regions produce antifreeze glycoproteins (AFGPs) or antifreeze proteins (AFPs) in their sera to protect them from freezing. This protein lowers the freezing temperature of solution without altering its melting temperature.

Thermal hysteresis, the difference between the freezing and melting temperature, is a unique property of these proteins. AFPs and AFGP have been demonstrated to bind to ice crystals and inhibit ice crystal growth.

Despite their similar antifreeze properties, these proteins are quite different in their protein structures. There are one type of AFGP and three types of AFP. Recently fourth type of AFP has also been reported in longhorn sculpin.

The Atlantic salmon Salmo salar lacks any of these AGFPs or AFPs gene(s) and are unable to survive in sub-zero sea water temperature. An inability to tolerate temperature below – 0.6 °C to – 0.80 °C is one of the major problems of sea cage farming in Northern Atlantic coast. Hew and his co-workers developed antifreeze-resistant Atlantic salmon containing the AFP or AFGP genes using gene transfer technology.

They used genomic clone (2A-7) encoding the major liver-type AFP (wflAFP-6, previously known as (HPLC-6) from the winter flounder (Pleuronectus amaricanus) was used as a candidate for gene transfer. Flounder AFPs belonged to the type I AFPs that are small polypeptides and high in alanine and helical content. Flounder AFPs is multi-gene family of 80-100 copies encoding two different isoforms, namely the liver type and skin type AFPs.

The liver type AFPs such as wflAFP-6 or wflAFP-8 (HPLC-8), are synthesized exclusively in the liver as prepro AFPs. In contrast the skin-type AFPs, including wfsAFP-2 and wfsAFP-3, are expressed widely in many peripheral tissues as intracellular mature AFPs.

Growth Hormone Gene Transfer

Recently scientists have developed an "all fish" growth hormone model. They have cloned and sequenced the grass carp and common carp carbonic anhydrase (CA) gene and growth hormone gene. The grass CA gene (beta-actin) promoter has been linked to a grass carp growth hormone cDNA to form a high efficiency expression vector called pCAZ. Using the CAT gene as receptor gene, a pCA grass carp growth hormone was microinjected into fertilized, non-activated common carp via the micropyle, generating "all fish" transgenic carp. The presence of transgene was detected by reverse transcriptase PCR and Northern blotting. These transgenic fish showed about 137% high growth rate of the control.

Disease-resistance Gene Transfer

In China scientists piloted a gene contributing resistance to the grass carp haemorrhagic virus (GCHV). Eleven different gene fragments encoding protein was cloned and isolated from translation in vitro using GCHV genomic single gene fragments. Based on the information of capsid protein SP6 and SP7 gene cDNA, 3 oligonucleotides were synthesized and fused with SV40 MT promoter and transferred into grass carp cytokine-induced killer (CIK) cells via a constructed expression vector and transfected with GCHV. The result indicated that the mortalities were reduced by one order after challenge with the virus.

Applications of Transgenic Fish

Transgenic Fish may be better used for the following purposes:

- For increasing fish production to meet the growing due to demand of food due to increase in world population.
- For production of pharmaceutical and other industrial products from piscine origin.
- For development of transgenic native glow fish varieties for aquarium.
- As fish biosensors for monitoring aquatic pollution.

- For isolation of genes, promoters and synthesis of effective gene constructs.
- For researches in embryonic stem cells and in-vitro embryo production.
- For production of anti-freeze protein.

Environmental Concerns about Transgenic Fish

The primary environmental concerns about releases of transgenic fish, for example, include competition with wild populations, movement of the transgene into the wild gene pool, and ecological disruptions due to changes in prey and other niche requirements in the transgenic variety versus the wild populations.

Transgenic Fish could Threaten Wild Populations

West Lafayette, Ind. — Purdue University researchers have found that releasing a transgenic fish to the wild could damage native populations even to the point of extinction. Transgenic fish could present a significant threat to native wildlife.

"Transgenic fish are typically larger than the native stock, and that can confer an advantage in attracting mates", Muir says. "If, as in our experiments, the genetic change also reduces the offspring's ability to survive, a transgenic animal could bring a wild population to extinction in 40 generations". Although at Canadian research facilities, elaborate precautions are being taken to prevent the release of transgenic fish into the environment. The fish are often raised in ponds covered with nets to keep birds out; enclosed by electric fences to keep muskrats, raccoons, and humans out; and the outlets are fitted with screened drains to prevent the loss of small fishes or eggs.

Gene Flow

One of the larger environmental concerns raised by transgenic fish is the possibility that a transgenic species raised in open water pens will escape and spread novel traits into the ecosystem by breeding with wild relatives, a biological process known as "gene flow." Gene flow between transgenic or conventionally bred fish and wild populations is an environmental concern, because it may present a threat to natural biodiversity. Some researchers believe that the genetic differences introduced to a transgenic fish may impact its net fitness, a scientific term meaning an organism's ability to survive and pass its genes to future generations.

The concept, which factors in characteristics such as the juvenile and adult viability of a fish, the number of eggs produced by a female, and the age at which a fish reaches sexual maturity, provides a useful barometer for discussing some gene flow scenarios. According to one scientific model, if a transgenic fish escapes and mates with a wild fish, gene flow could follow one of three scenarios.

Purge Scenario

When the net fitness of a transgenic fish is lower than that of its wild relatives, natural selection will quickly purge from the wild population any novel gene(s) introduced by the transgenic fish. In theory, evidence of the novel trait will disappear from subsequent generations.

Spread Scenario

When the net fitness of a transgenic fish is equal to or higher than the net fitness of a wild mate, gene flow is likely to occur and the genes of the transgenic fish will spread through the wild population. This means evidence of the transgenic genome would persist in subsequent generations.

Trojan Gene Scenario

When the net fitness of a transgenic fish is altered such that the fish has enhanced mating success but reduced adult viability (i.e., chances of surviving long enough to mate), introduction of that fish into the wild population could result in a rapid decline of the wild population. Essentially mating success would ensure the spread of the novel gene throughout the population, but the inability to survive would reduce the population size of subsequent generations and potentially lead to extinction. A declining fish population would also have secondary impacts on other aquatic species that feed on, or otherwise depend on it. Populations unable to' successfully switch over to another food source, or those whose survival or reproduction depends directly on the declining population, would also suffer.

Transgenic Fish Invasive Species

Even if they do not breed with wild relatives, transgenic fishes that escape into natural ecosystems could be an environmental nuisance by becoming an invasive species. This danger mainly arises for those transgenic fishes endowed with new genes that improve such fitness traits as breeding capabilities and the ability to withstand harsh conditions. The establishment of a thriving transgenic fish population in an ecosystem where it has never existed could crowd out native fish populations.

Risk Mitigation

It is important to note that developers of transgenic fish are attempting to reduce or eliminate both gene flow and invasive species risks by sterilizing transgenic fish. Sterilization is relatively easy and inexpensive but success rates are highly variable. In addition, sterilization does not necessarily neutralize environmental risks. Academic scientists note that an escaped, sterile fish might still engage in courtship and spawning behaviour, disrupting breeding in wild populations. Waves of escaped sterile fish could also create ecological disruptions as each group is replaced by another equally strong group of transgenic sterile fish.

Transgenic Pigs

The domestic swine genome has been altered through selective breeding to produce pigs with desired characteristics in agriculture for many centuries. The first "nontraditional" genetic modification in pigs was accomplished through microinjection of hundreds of copies of a foreign DNA fragment into the pronuclei of pig zygotes. The stable integration of a non-native gene into the swine genome resulted in the generation of the first transgenic pigs. This ability to insert new genetic material and/or subsequently delete or replace genes opened new possibilities for the use of pigs as a research animal. The medical community had already accepted pigs as an excellent model for surgical testing based on their respective organs' similarity to human heart, coronary vasculature, liver, kidney, lungs, and uterine histology. The prospect of obtaining genetically modified pigs further extended their biomedical potential, especially to mimic inherited human diseases and to establish experimental DNA-altering treatments (i.e. gene therapy). Pigs are ideal for this type of genetic research as the size and composition of the porcine genome is similar to that of humans. The completed draft of the pig genome will be based on the most recent assembly (*Sus scrofa* genome build 10 (Sscrofa10); currently being annotated), which represents about 98% of the porcine genome. The use of organs from transgenic pigs for xenotransplantation into humans and the production of pharmaceuticals are also being explored. From an agricultural standpoint, the worldwide demand for pork continues to rise as the populations and incomes of developing countries increase. Genetically altered swine have the potential to meet this demand by enabling development of strains with improved production traits, more efficient feed usage, increased reproduction, and greater protection from acquiring and transmitting disease.

Production of Transgenic Pigs

The first method to produce transgenic pigs, *pronuclear microinjection*, was originally established in mice. DNA is microinjected into the pronuclei of zygotes collected from a superovulated female, and then transferred to a recipient pig by embryo transfer. The technique is reliable and widely used, but only about 1% of injected eggs produce transgenic pigs. An alternative method, *oocyte transduction*, uses a replication deficient virus to deliver the transgene to the porcine oocytes and can increase the percentage of transgenic offspring. *Sperm-mediated DNA transfer* has also been used to produce genetically modified pigs with high efficiency based on the ability of sperm cells to internalize and integrate exogenous DNA during fertilization. Each of these methods can successfully make transgenic swine, but limitations include the inability to prescreen embryos for transgene integration prior to embryo transfer, the lack of expression specificity arising from random integration of foreign DNA, and the fact that only transgene addition is permitted, not deletion (i.e. gene knockout).

Targeted introduction of transgenes and loss-of-function mutations via homologous recombination in embryonic stem (ES) cells has been used for genetic manipulation of mice for decades. The lack of a stable source of true pig ES cells, however, has impeded the use of this method for creating chimeric, transgenic pigs with germline cells that participate in gamete formation (germilne transmission). One novel alternative approach is to graft transgenic male pig germ cells into immunodeficient mice (ectopic xenograft), resulting in the production of fully functional xenogeneic sperm for fertilization after intracytoplasmic sperm injection. This method is still in development, but viable non-transgenic piglets produced using sperm from ectopic testicular xenografts have been reported. It remains to be seen how successful transgenic, ectopic xenografts will be in the production of genetically modified pigs.

The most popular method of producing genetically modified pigs to date is through genomic modification of somatic cells followed by nuclear transfer (NT),. In this process, the nuclei of somatic cells are transferred into enucleated metaphase II oocytes, and then this complex is activated by electrofusion. Reconstructed embryos are then cultured and transferred to synchronized recipients for gestation. The advantages of NT for gene transfer in pigs was described by Robl and First, and nuclei from porcine blastomeres were used to produce the first cloned pigs from embryonic donor cells. Soon after the demonstration that the nuclei of adult somatic cells could undergo proper reprogramming to produce viable mammalian offspring, the first cloned pig derived from differentiated cells (porcine fetal fibroblasts) was reported and was soon followed by the cloning of pigs from cultured adult granulosa cells.

The combination of genetically modified somatic pig cells and NT technology has produced many valuable biomedical and agricultural swine models. Despite these achievements, NT remains an inefficient process in pigs, as it does in all mammals without stably cultured ES cells. Many reconstructed pig embryos fail early in pregnancy, requiring large numbers of embryo transfers per recipient. The prevailing view is that incomplete epigenetic reprogramming of donor cell nuclei results in aberrant gene expression during development. Embryo culture conditions that alter epigenetic modifications of DNA, including histone deacetylase inhibitors (e.g. Trichostatin A, Scriptaid), can have a beneficial effect on the development of cloned pig embryos. Pregnancy rates in recipients may also be improved by treating reconstructed embryos with the proteasome inhibitor MG132 after electrical fusion/activation. Transient treatment with MG132 likely prevents premature oocyte activation and lowers the rate of embryo fragmentation after NT. Found that by transferring cloned embryos immediately after activation to the oviducts of recipients that were 24 h asynchronous to the embryos increased pregnancy rate (9% with synchronous recipients versus 53% with asynchronous recipients) and survival of porcine embryos (1.8% vs. 12.5%). The researcher hypothesize that this increased cloning success is related to the freshly reconstructed embryos having more time for proper genetic reprogramming. Additional optimized methods of swine genetic modification and transfection have reduced donor cell culture times in an effort to improve NT offspring viability. After transfer to recipients, inadequate placental development is frequently observed in NT-derived pigs. Patterns of gene and protein expression in the extraembryonic membranes of NT pigs indicate improper development at the fetal-maternal interface, observations that imply a need for pharmacological interventions to improve development of the extraembryonic tissue. These advances have improved the ability to genetically modify porcine donor fibroblasts to produce cloned transgenic pigs to a state that a number of useful NT disease models exist, and are poised to have significant impacts in biomedicine and agriculture.

Biomedical Applications

Cell Tracking

Some of the first “proof-of-principle” models of transgenic pigs included NT clones that expressed enhanced green fluorescent protein (eGFP). Since that time, pigs that express red fluorescent protein and multi-gene fluorescent proteins (blue, green and red) have been produced. Tissues from these transgenic swine are useful in biomedical research requiring genetically and phenotypically marked cells or organs. In one such study, cells isolated from fetal eGFP-transgenic pigs were used to evaluate the survival of porcine retinal progenitor cells as allografts in wild-type recipient pigs with damaged retinas. Following transplantation, eGFP expression allowed histological visualization of cell integration in the recipient pigs, demonstrating the utility of these transgenic pigs as a large animal model.

Pig-to-Human Xenotransplantation

The first landmark swine model derived specifically for human health was in the field of xenotransplantation. The urgent need for replacement organs is unmet by human organ donations, with over 110,460 recipient candidates awaiting organ transplants in the United States. This has spurred a large scientific effort to make pig organs transplantable to humans. Rejection of pig organs by human recipients is caused by humoral and cell-mediated responses. Hyperacute rejection (HAR) occurs within 24 hours of transplantation and results from antibody recognition of galactose α1,3-galactose (α-Gal; epitopes synthesized by the enzyme α-1,3-galactosyltransferase (GGTA1). Acute humoral xenograft rejection (AHXR), usually begins after the first week of transplantation, and is caused by mismatched human leukocyte antigen (HLA) complex antigens that are present on all cells. To circumvent HAR, swine fibroblasts were produced with a knockout of the gene GGTA1. Production of cloned swine from these fibroblasts was a major victory in the search for human organ replacements. To augment protection against HAR, cloned GGTA1 knockout swine were produced with a transgene overexpressing α1,2-fucosyltransferase, which uses the same substrate that produces α-Gal, but instead produces a universally tolerated human antigen. Antibody-initiated complement activation is one dominant mechanism through which AHRX occurs. Transgenic cloned swine that overexpress the complement regulatory protein human-decay accelerating factor (hDAF) and an enzyme that modifies α-Gal, N-acetylglucosaminyltransferase III (GnT-III) have been developed to protect against AHXR and HAR, respectively. Swine with multiple genetic modifications (e.g. GGTA1 knockout combined with hDAF overexpression) appear to be the most promising route to widespread availability of organs for human transplant.

All pig breeds carry porcine endogenous retroviruses (PERVs) due to their integration in the swine genome. Although no known swine disease has been definitively attributed to PERVs, elevated viral expression in melanomas and pulmonary metastasis-derived cell cultures from selectively bred Munich miniature swine has been observed. The function of the virus in tumour development is still unclear. The greater concern for PERVs is the risk of infection of human cells during xenotransplantation of pig organs to humans, presenting a major hurdle to the use of pig organs in human recipients. To inhibit infection of human cells with PERVs, Dieckhoff et al. transfected porcine fibroblasts with a lentiviral vector expressing a short hairpin RNA (shRNA). The expression of a specific shRNA initiates the RNA interference (RNAi) cascade, prompting a significant inhibition of PERV expression in cloned piglets. If this level of PERV inhibition is sufficient to prevent human infection, these pigs may be combined with other genetically modifications described previously to advance successful organ xenotransplantation.

Cardiovascular Disease

Swine are historically considered an excellent model for the human cardiovascular system. As a potential dietary source of essential and beneficial fatty acids, genetically modified pigs have been developed to express Δ12 fatty acid desaturase (FAD2) from spinach to increase linoleic acid, as well as pigs expressing a humanized *Caenorhabditis elegans* gene, fat-1, encoding an n-3 fatty acid desaturase to increase the n-3:n-6 fatty acid ratio in meat. These pig models can also be used to examine the cardiovascular effects of an altered n-3:n-6 fatty acid ratio in the swine themselves. Another important regulator of vascular health is the signaling molecule, nitric oxide (NO). Transgenic swine that overexpress endothelial nitric oxide synthase, responsible for producing

NO in the inner lining of blood vessels, will increase our understanding of the complex regulation of vasodilation and potential therapies for diseases related to endothelial dysfunction. Vascular signaling by NO is closely associated with endothelial hydrogen peroxide (H_2O_2), a molecule that plays key roles in cardiovascular regulation and age-accelerated vascular disorders. To further explore the vascular role of H_2O_2, transgenic Yucatan minipigs that overexpress human catalase in the endothelium were developed. These large-animal cardiovascular models will enable real-time measurement of functional parameters including blood flow, temperature, tissue oxygenation, perfusion and diffusion that are difficult or impossible to monitor in similar genetically modified rodent models.

Cystic Fibrosis

Cystic fibrosis (CF) is an autosomal recessive disease affecting about 30,000 individuals in the United States. CF is caused by mutations in the gene that encodes for the cystic fibrosis transmembrane conductance regulator (CFTR) anion channel. Symptoms of CF arise from faulty water and ion transport across epithelial tissues, leading to inadequate hydration of mucous secretions. These thick secretions in organs such as the lungs initiate progressive cycles of infection and inflammation, the main cause of morbidity and mortality. Over 1600 mutations related to CF have been described, and although several CFTR knockout mouse models have been developed, none of them develop the obstructive lung disease symptoms seen in humans. In an effort to produce an accurate CF model, pigs were cloned from fetal fibroblasts in which the CFTR gene was mutated to produce null and ΔF508 (the most common genetic mutation in CF) alleles. These newborn CF pigs display defective chloride transport, meconium ileus, pancreatic destruction, and early focal biliary cirrhosis that occur in newborn humans with CF. In contrast to CF mouse models; these CF pigs develop the principal characteristics of human CF lung disease. Recently, the phenotype of CF pigs resolved a long-misunderstood question about the order of events in the inflammation/infection pathogenic cycle. The initial event displayed in CF pigs is impaired bacterial elimination from the lungs, followed by a cascade of inflammation and pathology. This key finding about CF pathogenesis provides new options for CF therapy and prevention.

Alzheimer's Disease

Alzheimer's disease (AD) is a progressive manifestation of dementia that typically begins with subtle and poorly recognized failure of memory, gradually becoming more severe, and eventually, incapacitating the patient. The genetic basis for familial, autosomal dominant AD (FAD) lays in mutations to the genes PSEN1, PSEN2, and the amyloid precursor protein gene (APP). These mutations are associated with increased production of proteolytic fragment Aβ which aggregates into fibrils and toxic oligomeric forms, initiating synaptic damage and neurodegeneration. A genetically modified pig model for FAD was developed based on similarities in size, physical features, and rate of growth between the porcine and human brain. Porcine fibroblasts hemizygous (carrying a single copy) for a neuron-specific splice variant of human APP were generated via transfection with a cDNA construct, followed by NT cloning. This splice-variant carries an AD-causing dominant mutation known as the Swedish mutation. The transgene was expressed in the brain, but the researcher speculates that it may take until the age of 1–2 years before Aβ accumulates to symptomatic levels in the porcine brain. Most (95%) AD cases of later onset, however, do not follow Mendelian inheritance, despite showing significant heritability. Pig models that address the genetic foundations of other forms of AD could provide translational data for early detection and new treatment in patients.

Spinal Muscular Atrophy

Spinal muscular atrophy (SMA) is an autosomal-recessive neurodegenerative disease, and the most common genetic cause of infant mortality. Clinical manifestations of SMA include motor neuron loss and skeletal muscle wasting. Spinal muscular atrophy (SMA) is caused by a deletion or mutation of the survival motor neuron (SMN1) gene. A highly conserved gene (SMN2) is also present exclusively in humans. Mutations in SMN2 have no clinical consequence if SMN1 is retained. In the case where SMN1 is mutated, the disease severity correlates inversely with SMN2 gene copy number, although SMN2 alone cannot prevent the disease.

A recent porcine animal model for SMA was developed specifically to evaluate the efficacy, pharmacology, and toxicology of SMA therapeutics. The authors highlighted the importance of demonstrating the conservation of alternative splicing events of both human SMN1 and SMN2 in the pig in order to produce a successful transgenic model of SMA. The porcine model of human SMA is being produced in three stages: First, a knockout of the SMN allele produced $SMN^{+/-}$ pigs. Next, a human SMN2 transgene is being added to the pig genome. Finally, $SMN^{-/-}$; SMN2 pigs will be generated through breeding and a second round of nuclear transfer. Nuclear transfer of SMN targeted fetal fibroblasts produced healthy $SMN^{+/-}$ piglets that, like their human $SMN^{+/-}$ counterparts, are phenotypically normal. Efforts are underway to introduce the human SMN2 transgene into swine fetal fibroblasts, leading to completion of the pig model of SMA. Effective SMA Therapeutics does not currently exist, highlighting the value of a genetically modified pig model for this disease.

Diabetes

Another transgenic swine model for human disease is for the phenotype of diabetes mellitus. Diabetes mellitus (DM) is defined by the American Diabetes Association as a group of metabolic diseases characterized by hyperglycemia resulting from defects in insulin secretion, insulin action, or both. Type 1 DM results from autoimmune destruction of β-cells in the pancreas, while Type 2 DM is a generalized metabolic disorder resulting from reduced tissue sensitivity to insulin. Chronic hyperglycemia caused by absent or ineffective insulin is associated with long-term damage to organs, especially the eyes, kidney, nerves, heart and blood vessels. Type 2 DM represents more than 80% of the DM cases, and approximately 57 million Americans are classified as "prediabetic". Type 3 DM is a maturity-onset diabetes of the young (MODY3), which is an early-onset non-insulin-dependent type of diabetes characterized by autosomal dominant inheritance.

A transgenic pig model for Type 3 DM was produced based on a mutation in the human hepatocyte nuclear factor 1α (HNF-1α) gene. The mutated HNF-1α construct was delivered by ICSI-mediated gene transfer followed by NT. Although 22 live transgenic piglets were farrowed, close to 80% of these clones died within 10 days of birth. The role of transgenesis and the cloning procedure in this mortality could not be determined. The surviving piglets, however, exhibited a diabetic phenotype, including high blood glucose and poor insulin secretion from the Islets of Langerhans. Although pig models of diabetes have previously been generated by chemical destruction of the β-cells, these cloned pigs form the first transgenic model where pigs exhibit the pathophysiological characteristics of diabetes. More recently, transgenic pigs expressing an impaired glucose-dependent insulinotropic polypeptide (GIP) receptor were generated via lentiviral transgenesis. Normally, insulin secretion is induced by GIP resulting from hyperosmolarity of glucose in the duodenum. This pig model exhibited significantly reduced oral glucose tolerance, demonstrating an essential role of GIP for insulin secretion. Additionally, proliferation of β-cells and physiological expansion of β-cell mass were significantly reduced in these transgenic pigs, closely resembling the characteristic features of human type 2 diabetes.

Pharmaceutical Production

The high cost of purified proteins required to treat diseases like hemophilia limits treatment in most developed countries, and is prohibitive in developing nations. The predominant methods of isolation of proteins from human plasma donations or from transgenic proteins produced in mammalian cell culture are costly and inefficient. For instance, only 10% of factor VIII present in one liter of donated plasma can be isolated. As an alternative, pig mammary bioreactors would be an excellent system for large-scale production of therapeutic proteins due to the high volume of milk that can be harvested per day (2–3 liters) and the quality of post-translational processing of complex proteins. It is estimated that the at the amount of transgenic protein isolated per milliliter of porcine milk, only 60 transgenic pigs would be required to supply the *entire* amount of factor IX needed in the United States. To put this in perspective, to isolate the same quantity of protein, 1,200,000 liters of plasma would be required, the equivalent of approximately 1,600,000 plasma donations.

There are several reports of using pigs to produce human recombinant proteins, by either pronuclear injection or NT, such as human factor VIII and IX, hemoglobin, human protein C, human erythropoietin, human granulocyte-macrophage colony stimulating factor, and von Willebrand factor. For such proteins to be used as replacements in human patients, herds of transgenic swine must be isolated in pathogen-free facilities, and there must be routine testing of isolated protein lots for quality and purity.

Agricultural Applications

Meat production, disease resistance, and efficient use of feed are the areas of agriculture that stand to benefit most from genetically modified swine production. The use of genetically modified pigs as a source of higher quality, more efficiently produced food with reduced environmental impact tends to fall under greater governmental and public scrutiny than the biomedical models previously discussed. This is likely due to the fact that biomedical pig models are restricted to the laboratory, whereas transgenic pigs in agriculture would present direct entry of transgenic material into the human food chain. It is prudent to assess the safety of any product entering the marketplace for human consumption, but genetically modified crops and animals are held to a higher approval standard. It is agreed that the risks must be addressed, but the assessment should constitute a *true* risk-benefit analysis that considers the risk, if any, against the potential benefits and the cost to the consumer, the producer, the environment and to the animals themselves if the technology is *not* used.

Meat Production

Early work by Pursel et al. showed that genes for growth-related hormones (e.g. growth hormone, insulin-like growth factor) could be expressed in pigs produced by DNA microinjection. These transgenic pigs had markedly reduced subcutaneous fat and variable growth enhancement relative to control pigs, but also exhibited problems such as fatigue, gastric ulcers, and low libido (although fertile). These effects were suggested to be due to the growth hormone produced, and suggestions were made to improve the health of the pigs generated.

Myostatin (MSTN) is the only secreted protein demonstrated to negatively influence muscle mass in vivo. Targeting this gene could result in enhanced muscle growth in pigs, yielding more lean meat per animal. To date, no myostatin knockout pigs have been generated, but porcine fetal fibroblasts transfected with a MSTN targeting vector have been reported to have significantly lower expression of MSTN mRNA compared to control fibroblasts. This suggests the potential for cloned MSTN-knockout pigs derived from these fibroblasts to have more muscle growth than wild-type pigs.

Enhanced production can also be achieved by improving the milk quality and/or amount available to newborn pigs. Wheeler and colleagues produced transgenic sows that express bovine α-lactalbumin in the mammary gland, which leads to an increase in milk production. Weight gain was higher in piglets suckling from alpha-lactalbumin gilts as compared to piglets suckling from controls.

Disease and Stress Resistance

Model pigs with enhanced defenses against pathogenic or environmental stresses are few, at present, but genetic modification could improve survivability and ultimately meat quality. Under high heat conditions, finishing pigs have reduced growth performance and poorer carcass characteristics due to their limited thermoregulation capacity. To improve pig thermotolerance, transgenic pigs overexpressing porcine heat shock protein HSP70.2 were produced by pronuclear microinjection. The survival rate of primary fibroblast cells under 45°C culture conditions was significantly higher from the transgenic pig compared to the non-transgenic pig. The researchers speculate that transgenic pigs overexpressing HSP70 might be more resistant to summer temperatures, thereby reducing mortality and economic loss in animal production.

Feed Efficiency

Increasing the absorption of nutrients from pig diets has the dual benefit of efficient food usage while reducing manure-based environmental pollution. The majority of phosphorus in livestock grain feed is in the form of natural phytate, a more stable and complex phosphorus compound that pigs are not able to digest completely, leading to large quantities of phosphorus in the pig manure. To address these concerns, developed transgenic pigs by pronuclear microinjection that expresses a protease-resistant *Escherichia coli* phytase transgene in the pig salivary gland. The transgenic phytase functions under the low-pH conditions of the pig digestive tract. Variability in phytase activity among transgenic pig lines was attributed to the positional effect of the transgene insertion. This genetic modification resulted in essentially complete digestion of dietary phytate and reduced fecal phosphorus excretion by up to 75%. Transgenic model pigs such as this present a new biological approach for reducing waste/pollution in swine farming and decrease the need for supplemental nutrients to be added to feeds.

Applications and Ethical Concerns of Transgenic Technology

Medical Applications

Transgenic animals were mainly developed as disease model for humans for studying genetic basis of human and animal disease such as Cancer, Cystic fibrosis, Rheumatoid arthritis, Alzheimer's etc. It also contributed in understanding mechanisms of disease resistance in humans and animals and in the development of pharmaceutical drugs and product testing and/ or screening. Transgenic cows, sheep and goats are capable of providing nutritional supplements such as insulin, growth hormone and blood clotting factors. Transgenic animals also provide means for organ transplantation. Patients die every year due to lack of organs replacement (heart, liver, or kidney).Transgenic pigs can provide the transplanted organs needed to ease the shortfall.

Agricultural Applications

Transgenesis had made it possible to develop desired traits such as increased milk production, high growth rate etc. by selective breeding in animals in a shorter time with more precision. It also increases the yield production of desired characteristics. HERMAN, a transgenic bull, was developed that carries a human Lactoferrin gene (increases iron content). Transgenic pigs and cattle weigh approximately double than normal animals and are used for meat production. Transgenic sheep grow heavy wool on them. Scientists are working on development of disease resistant animals such as influenza-resistant pigs, but this phase is limited to certain genes.

Industrial Applications

The extraction of polymer strands from the milk helped the scientists to create light material with toughness and flexibility that are used further for weaving military uniforms, medical micro sutures, tennis racket strings etc. Also, wide varieties of enzymes and micro-organisms are engineered through transgenesis, which are also capable of producing enzymes that enhance rate of industrial chemical reactions.

Ethical Concerns

Ethical concerns deals include both intrinsic and extrinsic objections. Intrinsic objections based heavily on emotions, nevertheless, will not disappear from public belief. Extrinsic objections include the purpose and consequences of transgenic animal production. The main application of transgenesis is to develop improve animal welfare outcomes. GMOs have been intentionally designed to suit human purposes but use of animals in science had always been controversial issue. Some of the progresses are considered morally unacceptable e.g. cloning. There are some moral or ethical factors concerned to genetic manipulation based on principle of 3 R's i.e. Reduction of animal numbers, refinement of practices and husbandry to minimize pain and distress and

Replacement of animals with non – animal alternatives wherever possible. Ethical concern also focuses in animal welfare concerns, that is:

- Use of animals in research causes great suffering to the animals.
- Each biological species has a right to exist as a separate identifiable entity.
- Due to experimentation, it had diminished the sense of identity and individuality.

Xenotransplantation Issues – transplantation or organs from transgenic pigs to humans can spread Zoonotic disease such as Mad Cow disease, which could have devastating consequences.

4

An Integrated Study of Stem Cells

Stem cells are the cells which have the potential to differentiate into several specific cell types. Stem cell technology involves the use of these stem cells for the treatment of a wide variety of diseases. Stem cells are broadly divided into totipotent stem cells, pluripotent stem cells and mesenchymal stem cells. The topics elaborated in this chapter will help in gaining a better perspective about stem cell technology and stem cell therapy.

Stem cells are cells that have the potential to develop into many different or specialized cell types. Stem cells can be thought of as primitive, "unspecialized" cells that are able to divide and become specialized cells of the body such as liver cells, muscle cells, blood cells, and other cells with specific functions. Stem cells are referred to as "undifferentiated" cells because they have not yet committed to a developmental path that will form a specific tissue or organ. The process of changing into a specific cell type is known as differentiation. In some areas of the body, stem cells divide regularly to renew and repair the existing tissue. The bone marrow and gastrointestinal tract are examples of areas in which stem cells function to renew and repair tissue.

The best and most readily understood example of a stem cell in humans is that of the fertilized egg, or zygote. A zygote is a single cell that is formed by the union of a sperm and ovum. The sperm and the ovum each carry half of the genetic material required to form a new individual. Once that single cell or zygote starts dividing, it is known as an embryo. One cell becomes two, two become four, four become eight, eight become sixteen, and so on, doubling rapidly until it ultimately grows into an entire sophisticated organism composed of many different kinds of specialized cells. That organism, a person, is an immensely complicated structure consisting of many, many, billions of cells with functions as diverse as those of your eyes, your heart, your immune system, the color of your skin, your brain, etc. All of the specialized cells that make up these body systems are descendants of the original zygote, a stem cell with the potential to ultimately develop into all kinds of body cells. The cells of a zygote are totipotent, meaning that they have the capacity to develop into any type of cell in the body.

The process by which stem cells commit to become differentiated, or specialized, cells is complex and involves the regulation of gene expression. Research is on-going to further understand the molecular events and controls necessary for stem cells to become specialized cell types.

Sources of Stem Cells

There are two main sources of stem cells, embryos, which gives embryonic stem cells and adult tissue that gives adult stem cells. Embryonic stem cells can be taken out from inner cell mass of a blastula. These types of stem cells will be pluripotent in nature. Almost every tissue in our body contains adult stem cells which are multi-potent in nature. Apart from these umbilical cords blood is almost a very rich of stem cells.

Embryonic Stem Cells

Embryonic stem cells are pluripotent cells isolated from the inner cell mass of a blastocyst, the early mammalian embryo that implants into the uterus. Embryonic stem cells self-renew by dividing and can differentiate into any specialised cell of the body, but not extra-embryonic tissues such as the placenta.

The successful establishment of human embryonic stem cells (hESCs) in culture has raised unprecedented public interest and expectation of treating intractable diseases such as diabetes, spinal cord injuries, neurodegenerative and cardiovascular diseases. Much of this enthusiasm was predicated on the unlimited self-renewal capacity of hESCs and their remarkable plasticity in differentiating into every cell type in our body. These features presented the tantalizing possibility of an unlimited cell source in regenerative medicine to generate any tissues to replace injured or diseased tissues. However, translating the potential of hESC into therapies has been challenging. Although translation of hESC has been severely impeded by social and political constraints placed on hESC research through ethical and religious concerns over the destruction of viable blastocysts during hESC isolation, the main challenges have been safety and technical issues.

Challenges in ESC therapy

Overcoming Tumor Formation

The two defining characteristics of ESCs are: 1) their pluripotency, or the potential to differentiate into all cell types in the adult body; and 2) their unlimited self-renewal capacity, or the ability to remain in an undifferentiated state and divide indefinitely. For mESCs, pluripotency is often demonstrated by the production of mESC-derived animals through germline transmission by chimeras resulting from injection of the cells into blastocysts or through tetraploid complementation. In hESCs, proof of pluripotency has been limited to formation of teratomas or teratocarcinomas, which are tumors composed of randomly distributed tissues from the three primordial germ layers in immunologically incompetent mice. Karyotypically normal, low passage hESCs form benign teratomas that do not contain undifferentiated tissues and are less invasive while high passage hESCs which have become karyotypically abnormal give rise to highly invasive, malignant teratocarcinomas.

Pluripotency coupled with unlimited self-renewal not only define ESCs, they are also the main appeal of ESC as the cell source for regenerative medicine but at the same time, pose significant challenges to the transplantation of differentiated ESCs to replace injured or diseased tissues. The propensity of ESC to differentiate into teratomas necessitates the need to eliminate any residual ESCs in the differentiated cell preparation. There have been many strategies to eliminate residual ESCs or enhance the purity of differentiated ESC preparations. The use of heterologous selectable gene markers such as antibiotic resistance gene or fluorescent protein markers is generally not a strategy of choice as this could introduce potentially deleterious gene mutations. Most of the strategies centered around the use of endogenous markers that are unique or highly expressed on ESCs and not on their differentiated progeny. For example, SSEA-4 and TRA-1-60 which are highly expressed on hESCs have shown to be highly efficient in physically removing contaminating ESCs by magnetic or fluorescence-activated cell sorters (MACS or FACS). Another strategy exploits the flotation density of cell on discontinuous density gradients such as

Puresperm- or Percoll-based gradients. Using a relatively novel strategy, Choo et al. has raised antibodies against undifferentiated hESCs and identified an antibody that was cytotoxic against hESCs by oncosis. This antibody was an IgM that recognizes podocalyxin-like protein-1(PODXL). hESCs that were treated with mAB 84 did not form teratoma when transplanted into SCID mice even after 18-24 weeks. Therefore, there are viable technologies to remove or reduce residual hESCs in differentiated hESC preparation and mitigate the risk of teratoma formation in patients receiving hESC-based cell therapy.

Overcoming Immunorejection

Like all tissue transplants, hESC-based cell therapy will have to circumvent host immune rejection to engraft in the recipients. One proposed strategy was to establish ESC repositories with lines expressing the combinations of HLA molecules that are compatible with HLA haplotypes present in the population. Alternatively, the host's immune system could be manipulated to induce tolerance to foreign tissues by ablation of donor-reactive T cell in the thymus, generation of tolerogenic dendritic cells and induction of T_{reg} cells. However, with the development of induced pluripotent stem cell technology that makes the creation of "patient-specific" pluripotent cells containing the same genetic material as the recipient a highly viable and practical option, the issue of host rejection has become a non-issue.

The quest to create "patient-specific" pluripotent cells began with therapeutic cloning or somatic cell nuclear transfer (SCNT) where the diploid nucleus of a somatic cell was injected into a haploid enucleated egg to be reprogrammed by soluble factors in the host cell. Upon stimulation, the re-programmed cell divides to form a blastocyst with an inner cell mass that has identical nuclear genetic composition as the nucleus donor. Although this approach has worked to generate ESCs from different animals such as mice, rabbits, cats, sheep, cattle, pigs, goats and even primates, no hESC has been generated through this approach as it remains a highly inefficient process and the use of human oocytes is ethically controversial. ESCs generated through SCNT are in principle, heterogeneous in their genetic composition as they contain nuclear DNA of the nucleus donor and mitochondrial DNA of the egg donor. This raises the possibility that SCNT-derived ESCs could be rejected by the innate immune system of the host with which the ESCs share the same nuclear but not mitochondrial genetic material.

The breakthrough in creating "patient-specific" pluripotent cells was achieved when Yamanaka demonstrated that the introduction of transcription factors which regulate ESC self-renewal, including Oct3/4 and Sox2 was sufficient to reprogram somatic cells into ES-like cells. These induced pluripotent stem cells (iPSCs) are karyotypically normal with gene expression profiles highly similar to ESCs and can differentiate into cells of all three germ layers. Apart from being patient-specific, the major attraction of iPSCs lies in their derivation from somatic tissues and not from ethically contentious issues such as human oocytes or embryos. However, retroviral and lentiviral vectors were required to express the transcription factors for reprogramming of the somatic cells and this carries a risk of insertional mutagenesis. To circumvent the need for viral vectors, non-viral genetic modification approaches were developed. Recently iPSCs were obtained via a direct delivery of reprogramming factors into cells using poly-arginine protein transduction domains or mRNA, thereby circumventing any form of genetic manipulation. These improvements have essentially abrogated the issue of host/donor cell immune compatibility and considerably enhanced the prospects of generating patient-specific iPSCs for regenerative medicine. However, a recent study demonstrated that some hiPSC derivatives exhibit limited expansion capability, increased apoptosis and early cellular senescence as compared to their hESC-derived counterparts, raising doubts about the clinical value of this reprogramming technology. Also, it remains to be determined if the progeny of these cells, which are genetically identical to the reprogrammed cell, will trigger any immune response when reintegrated into the donor.

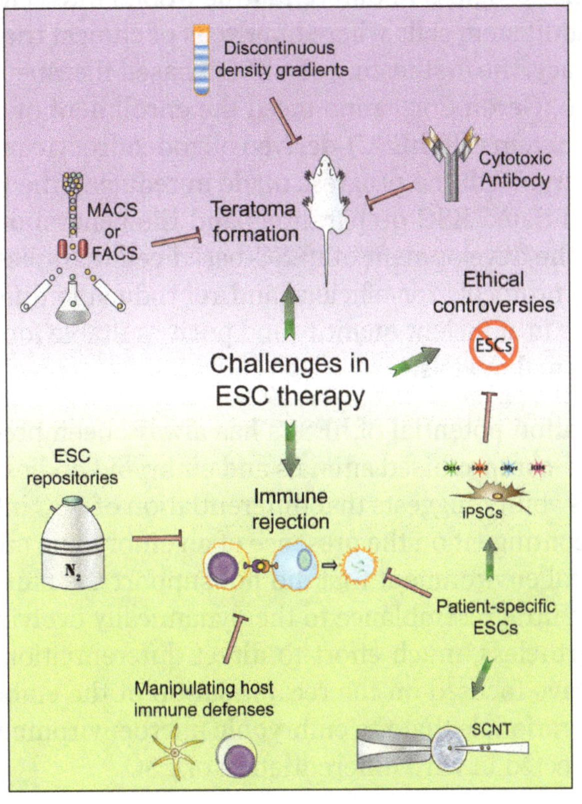

Figure shows mitigating tumor formation and immune rejection. Two of the major challenges to the translation of ESCs into clinical applications are teratoma formation by residual undifferentiated ESCs in the cell preparation and immune rejection of ESC-derived cells or tissues due to incompatible HLA profiles of ESC and recipient. To mitigate the risk of teratoma formation, several methods to remove residual hESCs have been developed using either physical or biological methods. Some of the physical separation methods are based on magnetic- or fluorescence-activated cell sorters (MACS or FACS) that sort against cells with ESC-associated surface markers, SSEA-4 and TRA-1-60 or on cellular density using discontinuous gradients of Percoll or PureSperm. Alternatively, residual ESCs can be destroyed using a cytotoxic antibody (mAb 84) specific for undifferentiated hESCs. To prevent immune rejection, one strategy proposed the establishment of ESC repositories to carry lines expressing HLA combinations compatible with all possible haplotypes in the population. Alternatively, donor cell tolerance can be induced by manipulating host immune defenses, such as eliminating donor-reactive T cells in the thymus, generating tolerogenic dendritic cells and inducing Treg cells. An ideal approach would be to generate patient-specific ESCs. Some of early efforts include the use of somatic cell nuclear transfer (SCNT). More recently, induced pluripotent stem cell (iPSC) technology has enabled with great ease the generation of self-pluripotent stem cells without the destruction of oocytes or embryos, hence bypassing ethical controversies.

ESC Differentiation

ESC owes its allure as the source of stem cells for regenerative medicine to two important potentials: 1) unlimited self-renewal potential and 2) the potential to differentiate into all the cell types in an adult. Unfortunately, the recent technological advances to circumvent the risks associated with transplantation of ESC-derived cells, namely teratoma formation and host immune rejection, were not matched by similar progress in differentiating hESCs into cells suitable for regenerative medicine. In contrast to adult stem cells where hundreds of clinical trials have been conducted to evaluate their clinical efficacy, the first testing of a hESC-based therapeutic candidate has only just been initiated. In Oct 2010, Geron Corp announced the enrollment of the first patient to test the safety of human embryonic stem cell (hESC)-derived oligodendrocyte progenitor cells, GRNOPC1, in treating spinal cord injury. With the progress made in reducing the risk of teratoma formation by residual ESC in differentiated ESC preparations and the generation of patient-specific iPSC, the major impediment to the development of hESC-based cell therapies remains the general lack of progress in developing protocols for efficient and reproducible differentiation of hESCs into clinically relevant cell types in sufficient quantity and purity suitable for transplantation studies in clinically relevant large animal models.

The pluripotent differentiation potential of hESCs has always been predicated on their ability to form teratomas in immune-compromised animals and embryoid bodies consisting of tissues from the three germ layers. This ability suggests that differentiation of ESC into the various cell types in the adult animal was not contingent on the presence of an embryonic microenvironment. Instead, it relies on a rather minimal environment that did not support the pluripotent and self-renewing state of the ESCs and bore little resemblance to the dynamically evolving microenvironment of a developing embryo. Nevertheless, much effort to direct differentiation of hESCs into potentially therapeutic cell types have focused on the recapitulation of the embryonic microenvironment based on a yet to be tested rationale that the embryonic microenvironment represents the optimal micro-environment for directed in vitro differentiation of ESC.

Recapitulating Embryonic Development to Induce Lineage Commitment

Embryogenesis is a highly dynamic complex process that is still being unraveled despite years of intensive research and much progress in elucidating the molecular and cellular processes involved in formation of an embryo. From a developmental perspective, the ESC represents cells that were frozen in the developmental state of a late-stage embryo just prior to differentiation and lineage commitment. The ability of ESC to re-enter the developmental process and differentiate when returned to the micro-environment of a blastocyst has provided compelling impetus to use the developing embryo to guide and direct in vitro differentiation of ESC to a specific cell type. Much effort has therefore been devoted to identifying the molecular cues that were involved in the differentiation of pluripotent cells in the blastocyst into specific terminally differentiated cells. The underlying rationale has always been that a temporal and spatial recapitulation of these cues in vitro will direct differentiation of ESC towards a specific cell type.

An early and critical phase of embryogenesis is gastrulation. During this process, the mono-layered blastula undergoes a series of transformation to form the tri-layered gastrula. The formation of these three germ layers (endoderm, mesoderm and ectoderm) marks the first stage of cell fate determination. This is followed by organogenesis when tissues and organs are formed from further differentiation of the germ layers. The endoderm gives rise to the epithelia of the gut and respiratory system, and organs such as liver and pancreas; the mesoderm gives rise to muscles, the circulatory system, bone and connective tissues; and the ectoderm gives rise to the nervous system and the epidermis. Similarly, the initial step towards deriving functional cells and tissues from ESCs may involve germ layer induction in vitro.

The first visible sign of gastrulation is the formation of the symmetry-breaking structure called the primitive streak (PS). Epiblast cells, which are derived from the inner cell mass, ingress through the PS to form the mesoderm and definitive endoderm. The remaining epiblast cells that do not ingress form the ectoderm. Many molecular factors have been implicated in this process and they include members of the large transforming growth factor β (TGFβ) and Wnt signaling families. Painstaking research has revealed some of the temporal and spatial effects of these factors during embryogenesis and many of these factors exerted similar effects on the differentiation of ESC cells. Differentiation of ESCs into each of the three germ layers could be induced by the same factors known to induce them during gastrulation. For example, Wnt, Nodal or BMP4 which have been shown to be important in the formation of epiblast cells in the PS of a developing embryo could similarly induce the formation of PS-like cells from ESC. As in gastrulation, exposure of the PS-like cells to high levels of Nodal further differentiate these cells to a Foxa2^{hi} cells that are comparable to cells in the anterior PS that forms the definitive endoderm. In contrast, exposure to Wnt, low level of activin (which activates Nodal) and BMP4 causes the PS-like cells to differentiate into a Flk-1^{+} posterior PS-equivalent population that forms the mesoderm. Therefore, the three germ layers can be induced in ESCs by exposing the cells to factors known to be important in the formation of these three germ layers during embryogenesis. Further, by modulating these factors in a concentration and temporal manner that recapitulates early embryonic development, commitment of ESCs to one of the germ layers could be enhanced.

Enhancing Lineage Commitment

The intensive research efforts to induce a bias in differentiating pluripotent ESCs towards one of the germ layers would, in principle, enhance the subsequent production of specific tissue cell types of this germ layer e.g. muscles from mesoderm. However, enhancing commitment of differentiating ESC to one of the three germ layers may not be the limiting factor in generating clinically useful cell types in sufficient number and purity for therapeutic or screening applications. For example, the most efficient derivation of clinically useful cell types from ESC is neural cell types and not surprisingly, the first ESC-derived cell type to be clinically tested is oligodendrocytes. The relative efficiency of generating neurons, astrocytes and oligodendrocytes from ESC probably lies not in the ease of generating neural progenitor cells but in the relatively high expansion capacity of ESC-derived neural progenitor cells. The high expansion capacity of neural progenitor cells would easily circumvent a limiting supply of rare neural progenitor cells formed during ESC differentiation and obviates the need to first bias differentiation of pluripotent ESCs towards an ectodermal germ lineage. Therefore, the rationale underlying the intensive research efforts to bias differentiating pluripotent ESCs towards one of the germ layers may be redundant at least for the derivation of neural cell types. Unlike ectodermal differentiation which is generally considered the default differentiation pathway for ESC, the derivation of mesodermal or endodermal cell types from ESC could still be enhanced by the recapitulation of early embryonic development processes to enhance mesodermal or endodermal commitment.

Terminal Differentiation of ESC

The use of a multi-stage protocol that attempts to temporally recapitulate embryonic development for the differentiation of hESC into insulin-producing pancreatic cells for diabetes treatment. During this differentiation regime, they observed the formation of sequential transient cell populations with markers that mapped onto the developmental pathway of pancreatic endoderm. The final cell population representing pancreatic endoderm was transplanted in mice for further differentiation and maturation. When these transplanted animals were treated with streptozotocin, the induction of hyperglycemia was attenuated. It was observed that some of the transplanted mice developed teratomas, suggesting the cell preparation was heterogenous and contaminated with ESCs that could differentiate into all cell types. In contrast, differentiation of mouse ESCs to insulin-producing cells is often a three-step protocol consisting of the formation of embryoid bodies, spontaneous differentiation into ecto-, endo- and mesoderm lineages and finally induction of pancreatic differentiation.

However, the cell populations generated using this protocol have low insulin content. Using an approach that combines elements from this protocol and that used in neural differentiation, we first derived highly expansible E-RoSH cell lines with meso-endoderm potential from spontaneously differentiating EBs. Like neural stem cells, these E-RoSH cell lines are highly proliferative and provide unlimited supply of cells for differentiation. Serum starvation and nicotinamide supplementation induce differentiation of E-RoSH cells to form a heterogenous, insulin-producing culture. Limiting dilution of such cultures yielded independently derived clonal insulin-producing ERoSHK cell lines. These cells contain equimolar of insulin and C-peptide that was stably maintained over 30 passages at a high concentration of 300-500 pmol/10^6 cells. The insulin-producing ERoSHK cells resemble pancreatic cells and display the defining functional properties of bona fide pancreatic beta cells. They synthesize and store insulin in typical intracellular vesicles. Under stimulation by secretagogues such as glucose, tolbutamide and glibenclamide, these cells close their ATP-sensitive K^+ channels, leading to membrane depolarization, opening of Ca^{2+} channels and the subsequent release of insulin and C-peptide in equimolar ratio, a mechanism resembling that of primary beta cells.

Most importantly, these cells can reverse hyperglycemia when grafted into streptozotocin-treated mice. Relative to their progenitor E-RoSH cells, ERoSHK cells also exhibit enhanced activity in biochemical pathways that are also highly characteristic of beta cells such as the pentose phosphate pathway, clathrin-mediated endocytosis and PPAR signaling. Importantly, transplantation of ERoSHK cells in hyperglycemic streptozotocin-treated mice reverses the hyperglycemia and removal of the transplanted cells restores the hyperglycemia. The transplanted cells do not form teratomas. Together, these studies illustrated the diversity of approaches that have been taken to differentiate ESCs to insulin-producing cells and the relative potential of each approach in generating the desired end product on a scale to support potential therapeutic application. They also prompted doubts on the need to recapitulate the precise developmental pathway when differentiating ESC.

From their perspective, developmental events directing duodenal endoderm towards an insulin-expressing β-cell phenotype are the result of millions of years of evolutionary selection, driven by environmental pressures rather than by conscious design. Therefore, instead of mapping experimental protocols on to the known developmental pathways of pancreatic endocrine cells, they proposed that conscious design may be a less circuitous route to arrive at the same end-point. However, in lieu of known developmental pathways, there is no obvious source to guide and rationalize such a design. In essence, a conscious design would inevitably have to be an empirical approach of careful observation, trial and error, and high throughput screens.

Empirical Differentiation of ESC

Despite a pervasive belief that a high fidelity recapitulation of developmental process represents the best strategy for efficient differentiation of pluripotent stem cells to therapeutically useful cell types, the two human ESC-derived cell types ready for testing in man were derived by empirically formulated protocols. Fortuitously, some elements in these protocols were subsequently found to map onto similar pathways in embryonic development.

Strategizing Differentiation of ESC for Therapeutic Applications

The progress of ESC-derived oligodendrocytes and RPE cells to clinical testing attests to the robustness and efficiency of the empirically-driven differentiation protocols. In contrast, differentiation of ESCs by meticulous mapping on embryonic development pathway has not yielded cells that are ready for clinical testing. Despite this dichotomy in outcomes, there is still a prevalent belief that high fidelity recapitulation of embryonic development process is the route to generate the most physiologically relevant cells. Embryonic development is a time tested success with defined milestones. In contrast, an empirically driven differentiation strategy is an inherently inefficient chance event.

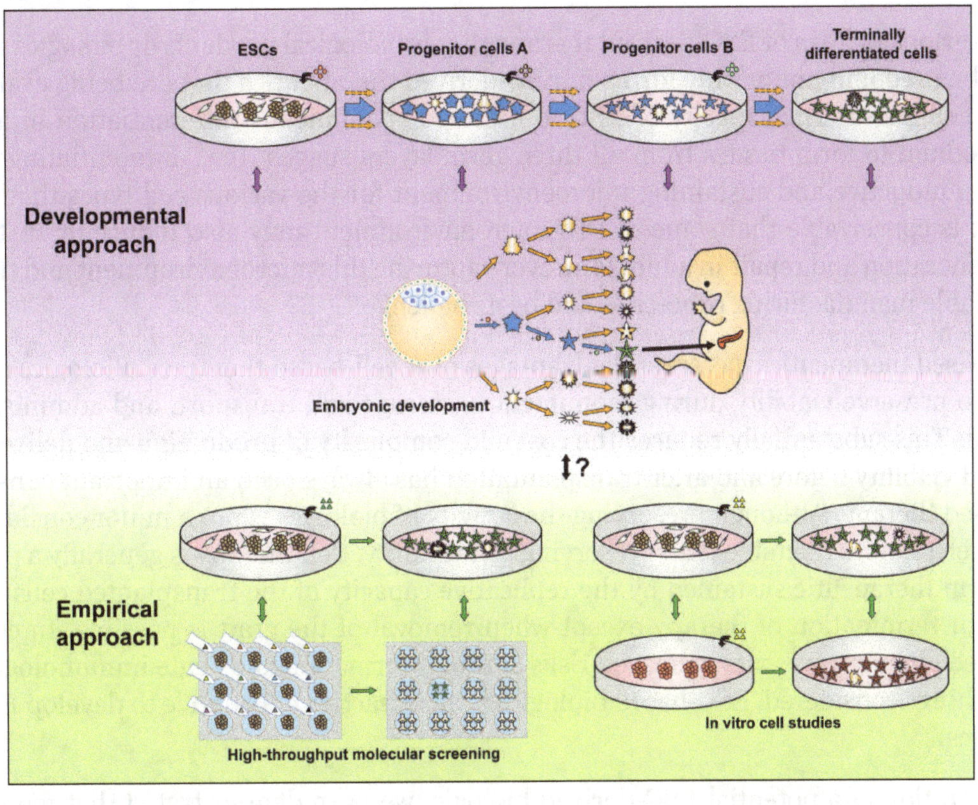

Figure shows strategies for differentiation of ESCs into therapeutically useful cell types. The strategies currently being used could be broadly classified into a developmental or an empirical approach. The developmental approach (upper panel) to produce a desired cell type (green stars) relies on the recapitulation of the developmental pathway (blue arrows) during embryogenesis that produces that desired cell type. The general expectation is that identifying the cues that direct the developmental process during embryogenesis and recapitulating these cues spatially and temporally in vitro will be most optimal in yielding physiologically functional cell type (e.g. pancreatic insulin-producing cells). The empirical approach involves the differentiation of ESCs either spontaneously, or using novel factors identified empirically, such as through high-throughput molecular screening or in vitro cell studies (e.g. neural induction of EC cells by RA). These factors may or may not play a role in development.

The success of a differentiation strategy based on embryonic development is predicated not only on the elucidation but also the recapitulation of the highly dynamic temporal and spatial changes in the embryonic microenvironment that is influenced by both intra- and extra-embryonic factors. Notwithstanding this, translating such a complex differentiation strategy to a scalable commercially viable manufacturing process will be an equally confounding unknown. On the other hand, developing a differentiation strategy using an empirical approach is a chance process of trial and error and fortuitous observation. This inherent inefficiency can be circumvented by high throughput screens to identify inducing molecules or combinations of molecules. There is also a likelihood that such a strategy would provide for a potentially scalable manufacturing process that will support clinical applications.

ESC Therapeutics: Cell Versus Biologic

A much overlooked form of ESC-derived therapeutics is biological products or biologics from ESC. To date, the predominant or only forms of ESC-derived therapeutics that are being evaluated are primarily cell-based. The capacity of ESC to undergo spontaneous differentiation in a minimal culture medium to form tissues from all three germ layers suggest that differentiating ESCs can produce an inductive and sustaining microenvironment for the various cell types that are being formed. It is conceivable that some of this microenvironment may also induce or sustain some tissue regeneration and repair in adult. However capturing this microenvironment and translating it to a scalable manufacturing process would be a challenge.

Biologic-based therapeutics has several advantages over cell-based therapies. Biologics eliminates the need to preserve viability during manufacture, storage and transport, and administration to the patient. This substantially reduces the cost and complexity of production and delivery. Maintaining cell viability before and after transplantation has always been an important consideration in cell-based therapy. Although preserving the activity of biologics is not a minor consideration, it is, nevertheless more tractable than preserving cell viability. Cell therapy is generally a permanent or long term therapeutic sustained by the replicative capacity of the transplanted cells with little recourse for termination of therapy except when removal of the graft is possible. In contrast to biologics, cell therapy presents increased risks of tumor formation and acute immunological rejections. All things considered, ESC-based biologics is an attractive alternative to develop ESC-based therapeutics.

As an illustration of a potential ESC-derived biologic, we have demonstrated that mesenchymal stem cells derived from hESCs secrete factors that are cardioprotective in pig and mouse models of myocardial ischemia/reperfusion injury. The active component in this secretion was small lipid vesicles of 50-100nm known as exosomes. Immortalization of these mesenchymal stem cells did not compromise the production or activity of the exosomes. These studies provided for the development of a sustainable scalable manufacturing process to produce potentially therapeutic exosomes for testing in the clinic.

Adult Stem Cells

Adult stem cells, like all stem cells, share at least two characteristics. First, they can make identical copies of themselves for long periods of time; this ability to proliferate is referred to as long-term self-renewal. Second, they can give rise to mature cell types that have characteristic morphologies (shapes) and specialized functions. Typically, stem cells generate an intermediate cell type or types before they achieve their fully differentiated state. The intermediate cell is called a precursor or progenitor cell. Progenitor or precursor cells in fetal or adult tissues are partly differentiated cells that divide and give rise to differentiated cells. Such cells are usually regarded as "committed" to differentiating along a particular cellular development pathway, although this characteristic may not be as definitive as once thought.

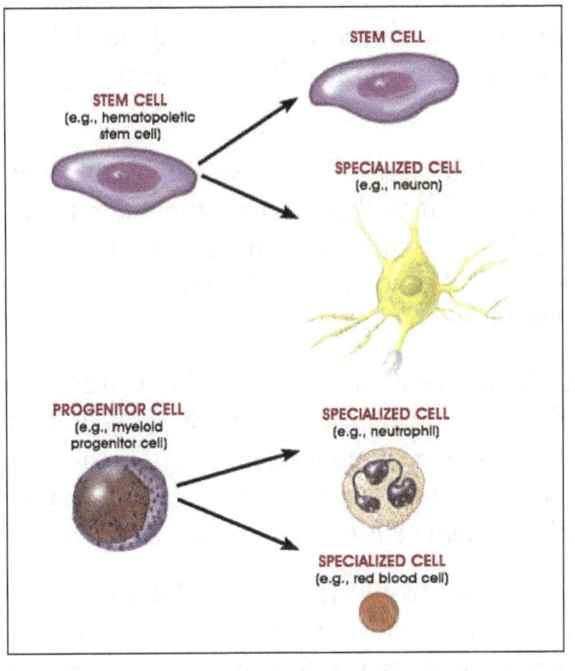

Figure shows distinguishing Features of Progenitor/Precursor Cells and Stem Cells. A stem cell is an unspecialized cell that is capable of replicating or self-renewing itself and developing into specialized cells of a variety of cell types. The product of a stem cell undergoing division is at least one additional stem cell that has the same capabilities of the originating cell. Shown here is an example of a hematopoietic stem cell producing a second generation stem cell and a neuron. A progenitor cell (also known as a precursor cell) is unspecialized or has partial characteristics of a specialized cell that is capable of undergoing cell division and yielding two specialized cells. Shown here is an example of a myeloid progenitor/precursor undergoing cell division to yield two specialized cells (a neutrophil and a red blood cell).

Adult stem cells are rare. Their primary functions are to maintain the steady state functioning of a cell—called homeostasis—and, with limitations, to replace cells that die because of injury or disease. For example, only an estimated 1 in 10,000 to 15,000 cells in the bone marrow is a hematopoietic (blood forming) stem cell (HSC). Furthermore, adult stem cells are dispersed in tissues throughout the mature animal and behave very differently, depending on their local environment. For example, HSCs are constantly being generated in the bone marrow where they differentiate into mature types of blood cells. Indeed, the primary role of HSCs is to replace blood cells. In contrast, stem cells in the small intestine are stationary, and are physically separated from the mature cell types they generate. Gut epithelial stem cells (or precursors) occur at the bases of crypts—deep invaginations between the mature, differentiated epithelial cells that line the lumen of the intestine. These epithelial crypt cells divide fairly often, but remain part of the stationary group of cells they generate.

Unlike embryonic stem cells, which are defined by their origin (the inner cell mass of the blastocyst), adult stem cells share no such definitive means of characterization. In fact, no one knows the origin of adult stem cells in any mature tissue. Some have proposed that stem cells are somehow set aside during fetal development and restrained from differentiating. Most of the information about adult stem cells comes from studies of mice. The list of adult tissues reported to contain stem cells is growing and includes bone marrow, peripheral blood, brain, spinal cord, dental pulp, blood vessels, skeletal muscle, epithelia of the skin and digestive system, cornea, retina, liver, and pancreas.

In order to be classified as an adult stem cell, the cell should be capable of self-renewal for the lifetime of the organism. This criterion, although fundamental to the nature of a stem cell, is difficult to prove in vivo. It is nearly impossible, in an organism as complex as a human, to design an experiment that will allow the fate of candidate adult stem cells to be identified in vivo and tracked over an individual's entire lifetime.

Ideally, adult stem cells should also be clonogenic. In other words, a single adult stem cell should be able to generate a line of genetically identical cells, which then gives rise to all the appropriate, differentiated cell types of the tissue in which it resides. Again, this property is difficult to demonstrate in vivo; in practice, scientists show either that a stem cell is clonogenic in vitro, or that a purified population of candidate stem cells can repopulate the tissue.

An adult stem cell should also be able to give rise to fully differentiated cells that have mature phenotypes, are fully integrated into the tissue, and are capable of specialized functions that are appropriate for the tissue. The term phenotype refers to all the observable characteristics of a cell (or organism); its shape (morphology); interactions with other cells and the non-cellular environment (also called the extracellular matrix); proteins that appear on the cell surface (surface markers); and the cell's behavior (e.g., secretion, contraction, synaptic transmission).

The majority of researchers who lay claim to having identified adult stem cells rely on two of these characteristics—appropriate cell morphology, and the demonstration that the resulting, differentiated cell types display surface markers that identify them as belonging to the tissue. Some studies demonstrate that the differentiated cells that are derived from adult stem cells are truly functional, and a few studies show that cells are integrated into the differentiated tissue in vivo and that they interact appropriately with neighboring cells. At present, there is, however, a paucity of research, with a few notable exceptions, in which researchers were able to conduct studies of genetically identical (clonal) stem cells. In order to fully characterize the regenerating and self-renewal capabilities of the adult stem cell, and therefore to truly harness its potential, it will be important to demonstrate that a single adult stem cell can, indeed, generate a line of genetically identical cells, which then gives rise to all the appropriate, differentiated cell types of the tissue in which it resides.

Evidence for the Presence of Adult Stem Cells

Adult stem cells have been identified in many animal and human tissues. In general, three methods are used to determine whether candidate adult stem cells give rise to specialized cells. Adult stem cells can be labeled in vivo and then they can be tracked. Candidate adult stem cells can also be isolated and labeled and then transplanted back into the organism to determine what becomes of them. Finally, candidate adult stem cells can be isolated; grown in vitro and manipulated, by adding growth factors or introducing genes that help determine what differentiated cells types they will yield. For example, currently, scientists believe that stem cells in the fetal and adult brain divide and give rise to more stem cells or to several types of precursor cells, which give rise to nerve cells (neurons), of which there are many types.

It is often difficult—if not impossible—to distinguish adult, tissue-specific stem cells from progenitor cells, which are found in fetal or adult tissues and are partly differentiated cells that divide and give rise to differentiated cells. These are cells found in many organs that are generally thought to be present to replace cells and maintain the integrity of the tissue. Progenitor cells give rise to certain types of cells—such as the blood cells known as T lymphocytes, B lymphocytes, and natural killer cells—but are not thought to be capable of developing into all the cell types of a tissue and as such are not truly stem cells. The current wave of excitement over the existence of stem cells in many adult tissues is perhaps fueling claims that progenitor or precursor cells in those tissues are instead stem cells. Thus, there are reports of endothelial progenitor cells, skeletal muscle stem cells, epithelial precursors in the skin and digestive system, as well as some reports of progenitors or stem cells in the pancreas and liver.

Adult Stem Cell Plasticity

It was not until recently that anyone seriously considered the possibility that stem cells in adult tissues could generate the specialized cell types of another type of tissue from which they normally reside—either a tissue derived from the same embryonic germ layer or from a different germ layer. For example, studies have shown that blood stem cells (derived from mesoderm) may be able to generate both skeletal muscle (also derived from mesoderm) and neurons (derived from ectoderm). Stem cells derived from one adult tissue can change their appearance and assume characteristics that resemble those of differentiated cells from other tissues.

Approaches for Demonstrating Adult Stem Cell Plasticity

To be able to claim that adult stem cells demonstrate plasticity, it is first important to show that a cell population exists in the starting tissue that has the identifying features of stem cells. Then, it is necessary to show that the adult stem cells give rise to cell types that normally occur in a different tissue. Neither of these criteria is easily met. Simply proving the existence of an adult stem cell population in a differentiated tissue is a laborious process. It requires that the candidate stem cells are shown to be self-renewing, and that they can give rise to the differentiated cell types that are characteristic of that tissue.

To show that the adult stem cells can generate other cell types requires them to be tracked in their new environment, whether it is in vitro or in vivo. In general, this has been accomplished by obtaining the stem cells from a mouse that has been genetically engineered to express a molecular tag in all its cells. It is then necessary to show that the labeled adult stem cells have adopted key structural and biochemical characteristics of the new tissue they are claimed to have generated. Ultimately—and most importantly—it is necessary to demonstrate that the cells can integrate into their new tissue environment, survive in the tissue, and function like the mature cells of the tissue.

In the experiments reported to date, adult stem cells may assume the characteristics of cells that have developed from the same primary germ layer or a different germ layer. For example, many plasticity experiments involve stem cells derived from bone marrow, which is a mesodermal derivative. The bone marrow stem cells may then differentiate into another mesodermally derived tissue such as skeletal muscle, cardiac muscle or liver.

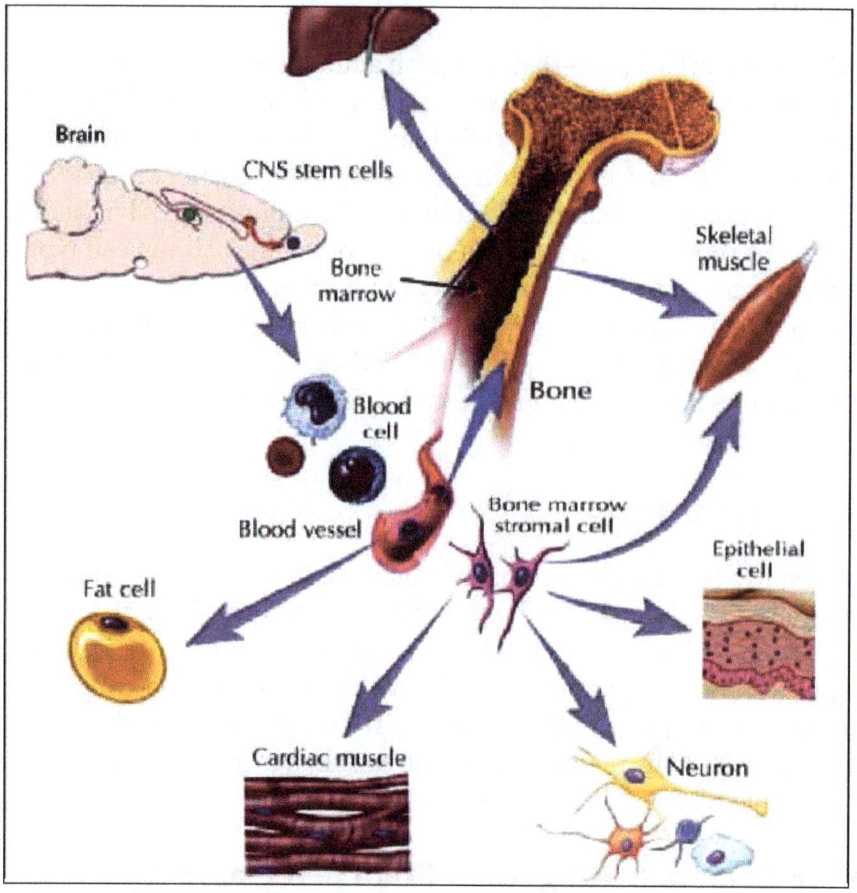

Figure: Preliminary Evidence of Plasticity Among Nonhuman Adult Stem Cells.

Alternatively, adult stem cells may differentiate into a tissue that—during normal embryonic development—would arise from a different germ layer. For example, bone marrow-derived cells may differentiate into neural tissue, which is derived from embryonic ectoderm. And—reciprocally—neural stem cell lines cultured from adult brain tissue may differentiate to form hematopoietic cells, or even give rise to many different cell types in a chimeric embryo. The cells would be deemed to show plasticity, but in the case of bone marrow stem cells generating brain cells, the finding is less predictable.

In order to study plasticity within and across germ layer lines, the researcher must be sure that he/she is using only one kind of adult stem cell. The vast majority of experiments on plasticity have been conducted with adult stem cells derived either from the bone marrow or the brain. The bone marrow-derived cells are sometimes sorted—using a panel of surface markers—into populations of hematopoietic stem cells or bone marrow stromal cells. The HSCs may be highly purified or partially purified, depending on the conditions used. Another way to separate population of bone marrow cells is by fractionation to yield cells that adhere to a growth substrate (stromal cells) or do not adhere (hematopoietic cells).

To study plasticity of stem cells derived from the brain, the researcher must overcome several problems. Stem cells from the central nervous system (CNS), unlike bone marrow cells, do not occur in a single, accessible location. Instead, they are scattered in three places, at least in rodent brain—the tissue around the lateral ventricles in the forebrain, a migratory pathway for the cells that leads from the ventricles to the olfactory bulbs, and the hippocampus. Many of the experiments with CNS stem cells involve the formation of neurospheres, round aggregates of cells that are sometimes clonally derived. But it is not possible to observe cells in the center of a neurosphere, so to study plasticity in vitro, the cells are usually dissociated and plated in monolayers. To study plasticity in vivo, the cells may be dissociated before injection into the circulatory system of the recipient animal, or injected as neurospheres.

Evidence for Plasticity

The differentiated cell types that result from plasticity are usually reported to have the morphological characteristics of the differentiated cells and to display their characteristic surface markers. In reports that transplanted adult stem cells show plasticity in vivo, the stem cells typically are shown to have integrated into a mature host tissue and assumed at least some of its characteristics. Many plasticity experiments involve injury to a particular tissue, which is intended to model a particular human disease or injury. However, there is limited evidence to date that such adult stem cells can generate mature, fully functional cells or that the cells have restored lost function in vivo. Most of the studies that show the plasticity of adult stem cells involve cells that are derived from the bone marrow or brain. To date, adult stem cells are best characterized in these two tissues, which may account for the greater number of plasticity studies based on bone marrow and brain. Collectively, studies on plasticity suggest that stem cell populations in adult mammals are not fixed entities, and that after exposure to a new environment, they may be able to populate other tissues and possibly differentiate into other cell types.

It is not yet possible to say whether plasticity occurs normally in vivo. Some scientists think it may, but as yet there is no evidence to prove it. Also, it is not yet clear to what extent plasticity can occur in experimental settings, and how—or whether—the phenomenon can be harnessed to generate tissues that may be useful for therapeutic transplantation. If the phenomenon of plasticity is to be used as a basis for generating tissue for transplantation, the techniques for doing it will need to be reproducible and reliable. In some cases, debate continues about observations that adult stem cells yield cells of tissue types different than those from which they were obtained.

Umbilical Cord Blood

Over the years, transplants of bone marrow stem cells have been used clinically to treat disease processes in which stem cells are beneficial. Umbilical cord blood stem cells were historically considered a waste product of the birthing process but are now known to have up to 10 times more stem cells than adult bone marrow.

Research on stem cell transplants began in the 1950s, with successful bone marrow transplants occurring in the 1970s, often to treat cancer patients whose own bone marrow was destroyed by chemotherapy and radiation. The first successful umbilical cord blood stem cell transplant was reported as occurring in the late 1980s. The recipient was a 6-year-old American boy from North Carolina who was treated for Fanconi's anemia (a genetic disorder) at Hospital St. Louis in Paris, France, using cord blood obtained from his younger sister's birth. Interestingly, more than 20 years after the transplant, this young man is alive and well. Not only did he survive long term, but both his immune system and his blood were transformed by the transplant of his sister's cord blood stem cells. Soon after this first documented cord blood stem cell transplant, the first public umbilical cord blood bank was established in 1991 in New York.

Prior to transplanting any type of tissue, a "matching" process must occur to increase the success of the transplant and decrease the likelihood that the transplant will be rejected. The rejection of a transplanted tissue is called "graft versus host disease." The matching process dates back to the late 1950s when the human leukocyte antigens were discovered. There are two classes of human leukocyte antigens. The first class is located on the surface of almost all of the cells with a nucleus within the body of the cell. The second class of human leukocyte antigens is located on the surface of immune cells. Each of the two classes of antigens has three subgroups, creating six antigens for which matching can occur. Thus, a "6 of 6" matching of the antigens represents a "perfect" match. Beyond the matching process, other factors contribute to the success or failure of a stem cell transplant. These factors include, but are not limited to, the age of both the donor and the patient, the type of disease being treated, and the number of stem cells being transplanted.

Uses of Stem Cell Transplants

Four main types of physical conditions are treated with stem cell transplants: cancers, blood disorders, congenital metabolic disorders, and immunodeficiencies. Examples of cancers that are treated with stem cells are both lymphoma and leukemia. Nonmalignant hemologic disorders also account for a fair share of the recipients of stem cells. Examples of these blood disorders are various types of anemias, such as sickle-cell anemia and Fanconi's anemia (the first disorder treated with umbilical cord blood stem cells). Stem cells have also been used to treat various metabolic disorders, such as adrenoleukodystrophy. The fourth major category of uses for stem cells is in treating immunodeficiencies, such as Duncan's disease or adenosine deaminase deficiency.

Table: Examples of Conditions Treated With Stem Cell Transplants.

Cancers	Blood Disorders	Congenital Metabolic Disorders	Immunodeficiencies
Acute lymphocytic leukemia Acute myelogenous leukemia Chronic myelogenous leukemia Myelodysplastic syndrome Neuroblastoma Hodgkin's disease Non–Hodgkin's lymphoma Burkitt's lymphoma	Sickle-cell anemia Fanconi's anemia Thalassemia Evan's syndrome Congenital cytopenia Aplastic anemia Diamond–Blackfan anemia Amegakaryocytic thrombocytopenia	Adrenoleukodystrophy Gunther's disease Gaucher's disease Hurler's syndrome Hunter's syndrome Krabbe's disease Sanfilippo's syndrome Tay-Sachs' disease	Adenosine deaminase deficiency Wiskott-Aldrich's syndrome Duncan's disease Ataxia-telangiectasia DiGeorge's syndrome Myelokathexis Hypogammaglobulinemia Severe combined immunodeficiency

Researchers continue to investigate new applications of stem cells. Ballen reported on studies examining the use of stem cells for treating autoimmune diseases, such as lupus, systemic sclerosis, and multiple sclerosis. Gunning reported on stem cell research for regenerative uses for heart attacks, stroke, spinal cord injury, diabetes, liver injury, and even traumatic brain injury. However, Gunning also noted that these regenerative uses for stem cells are purely in the research stage and, so far, no tangible evidence supports any clinical uses beyond the diseases that are currently being treated. Regenerative uses for stem cells are purely in the research stage and, so far, no tangible evidence supports any clinical uses beyond the diseases that are currently being treated.

Advantages of Umbilical Cord Blood Stem Cells

There are several advantages of using umbilical cord blood stem cells over bone marrow stem cells for transplants. The first advantage is that umbilical cord blood is relatively easy to collect and process. Once considered a substance to be thrown away after a birth, now the cord blood can be easily saved. After it is saved and sent to a storage facility, the cord blood is quickly available for use within days to weeks after processing. In contrast, bone marrow stem cells can take much longer to find a match, collect the sample, and process. The process for bone marrow transplantation can take from weeks to months. The collection process for cord blood is not painful to either mother or child and can be done either prior to or after the delivery of the placenta. Bone marrow transplants, on the other hand, require the donor to be hospitalized, anesthetized, and experience post collection pain and discomfort. Thus, compared to cord blood, bone marrow collection and transplantation of stem cells are more costly.

Advantages of Umbilical Cord Blood Stem Cells Versus Bone Marrow Stem Cells

- Ease of collection.
- No risk for mother or child.
- Less time needed for processing (more quickly available for use).
- Less costly than bone marrow collection.
- Less risk for transmission of infection.
- Less need for stringent antigen typing.
- Less rejection.

Another advantage of using umbilical cord blood stem cells is the decreased risk of the transmission of infectious disease. This particular advantage is partly because umbilical cord blood is almost never contaminated by Epstein-Barr virus or cytomegalovirus. Additionally, the processing of cord blood includes collecting data on the history of infection during the mother's pregnancy. For example, if the pregnant woman has a history of group B streptococcus, active genital herpes, or prolonged rupture of membranes and chorioamnionitis, umbilical cord blood is not saved. Generally, samples of the mother's blood are also drawn to test for infectious diseases, such as hepatitis, human immunodeficiency virus, and syphilis. Furthermore, after the cord blood units are collected, they are screened for disease, and any units that are deemed contaminated or infected are thrown away.

The stem cells obtained from umbilical cord blood are also less likely than bone marrow stem cells to be rejected in transplants. Considered to be immunologically immature, umbilical cord blood stem cells produce significantly fewer natural killer cells, creating a substantial decrease in rejection. Consequently, cord blood stem cells require less rigorous antigen tissue matching for transplants than bone marrow stem cells. Research indicates that a mismatch of up to two antigen sites still provides successful clinical outcomes. In fact, researchers report that the rate of rejection for cord blood stem cell transplants is half the rate of rejection for bone marrow transplants. When compared directly in cases of mismatched antigens, there was clearly less rejection in transplants involving cord blood stem cells than bone marrow stem cells.

Disadvantages of Umbilical Cord Blood Stem Cells

Despite the benefits of using umbilical cord blood stem cells for transplant, the process also has some disadvantages. For stem cell transplants to be successful, measurable signs of engraftment must occur. Engraftment is the opposite of rejection and indicates that the stem cell transplant is "working." Two measurable signs of engraftment are the recovery of both neutrophil (a type of white blood cell) and platelet (a clotting factor) production. These two clinical signs of recovery take longer to occur in umbilical cord blood stem cell transplants than in bone marrow stem cell transplants. In other words, the lab values for white blood cell production and platelet production take longer to increase after umbilical cord blood stem cell transplants than after bone marrow stem cell transplants. The disadvantages are:

- Slow engraftment.
- Limited cell dose:
    - Small volume of unit.
    - Additional cell doses unavailable.
- Autologous donation may have limited benefit owing to hereditary disorders.
- Storage issues:
    - Unknown length of long-term storage.
    - Cost related to long-term storage.
    - Quality control.

One of the factors that influence engraftment time is cell dose. Cell dose is directly related to the volume of umbilical cord blood collected. Cell dose refers to the amount of useful stem cells in the sample of blood. Because of the limited volume of cells collected from cord blood, the amount of stem cells in cord blood is approximately 10% less than the amount obtained from bone marrow.

A single unit of umbilical cord blood usually contains 50 to 200 ml of blood. If an amount of cord blood is less than this minimum volume, the unit is discarded as being unsatisfactory because the cell dose of the sample would not be high enough. Collecting an insufficient volume of cord blood occurs in about 50% or more cases of cord blood collection. In general, fewer stem cells are needed for cord blood transplantation, and usually a volume of 50 to 100 ml of cord blood will provide enough of a cell dose for a child or small adult. However, should the recipient need additional stem cells, it is impossible to obtain more stem cells from the infant because the cord blood volume is a limited amount?

Another important disadvantage that is not well understood by the general public is the limited use of an infant's own umbilical cord blood stem cells later in life, called an autologous transplant. Commercial cord blood banks often advertise the banking of the infant's cord blood as "biologic insurance." However, the chance that a child would be able to use his or her own cord blood is extremely small: from a 1:400 to a 1:200,000 chance over the child's lifetime. In fact, there are certain instances in which the use of one's own umbilical cord blood is contraindicated, as in cases when the defect is of a genetic origin. For example, autologous cord blood stem cells cannot be used to treat malignant cancers such as leukemia because the genetic mutations for the cancer already exist on the DNA of the cord blood. Using one's own stem cells would be, in effect, "contaminating" oneself with the same disease process. The chance that a child would be able to use his or her own cord blood is extremely small: from a 1:400 to a 1:200,000 chance over the child's lifetime.

Another important consideration for autologous use is that, currently, it is unknown how long umbilical cord blood will maintain its usefulness while frozen. Research indicates that cord blood stem cells can be maintained up to 15 years, but it is unknown if the cells would be preserved over the entire lifetime of a person. Furthermore, financial costs are associated with maintaining the cord blood over time. Kaimal, Smith, Laros, Caughey, and Cheng studied the cost-effectiveness of private umbilical cord blood banking for autologous use and concluded that it was not cost-effective in most instances because the chances that it would be used are extremely small.

Finally, there is a significant lack of regulation for umbilical cord blood banking. The lack of quality control, in turn, affects the quality of the specimen available for transplant. Some cord blood banks have submitted to voluntary accreditation, but the process of accreditation varies from bank to bank, whether public or private.

Types of Umbilical Cord Blood Banks

There are three types of umbilical cord blood banks: private, public, and direct-donation banks. The private bank is a commercial, for-profit entity that often advertises directly to expectant parents. These banks are designed for the sole use of the families who have saved the cord blood. Private Banks charge an initial fee for collection and processing and, then, a yearly fee to maintain the specimen. Another fee is often charged when a sample is removed for testing or treatment.

Public umbilical cord blood banks accept altruistic donations of cord blood and do not charge donation fees. Donated units are also processed, antigen typed, and frozen, ready for use. Unlike private banks, public banks do not reserve the units for the family that donated them; rather, units are available to the general public. In fact, a family that donates the blood would be no more likely to be a recipient of the blood than anyone else in the general population. Public cord blood banks function much like venous blood banks. The blood is released on an "as-needed" basis, and a processing fee may be charged to recoup some of the cost of storage.

Direct-donation umbilical cord blood banks function as an amalgamation of public and private banks. Direct-donation banks collect cord blood without charging fees. In addition, they accept autogenous donations and reserve them only for the family, especially for a family whose infant has a sibling with a disorder that may be treated with umbilical cord blood stem cells.

Types of Stem Cells

Totipotent Stem Cells

Totipotent stem cells are cells that have the capacity to self-renew by dividing and to develop into the three primary germ cell layers of the early embryo and into extra-embryonic tissues such as the placenta. A fertilised egg is a totipotent stem cell and as such can develop into any specialised cell found in the organism.

Totipotent stem cells are unique as they have a greater developmental potential compared with other stem cells. Having the ability to isolate and culture totipotent stem cells creates many therapeutic and research possibilities[1] related to:

- Studying zygotic genome activation (the point at which development becomes exclusively controlled by the zygotic genome, rather than the maternal genome).
- Rewiring the epigenome (treating pathological conditions induced by epigenetic alterations).
- Understanding early embryonic development in more detail.
- Creating human-animal chimeras[6] (theoretically, animals with human organs could aid disease modeling, drug development and transplantation).

Stem cells are increasingly being used as model systems in research. Differentiating between stem cell types relies on an understanding of the embryological roadmaps and the factors that define their pathways.

Source of Totipotent Stem Cells

The diploid zygote cell is totipotent. Totipotent cells also exist in subsequent divisions of the zygote, before the rise of the trophectoderm lineage (which occurs approximately four days after fertilization, depending on the species). To understand the origin of totipotent stem cells, it helps to be familiar with the early stages of embryonic development, and the related terminology. The table below highlights the important stages of early embryonic development and demonstrates the defining point at which totipotent stem cell no longer exists, and pluripotent cells arise.

Major Stages of Early Embryonic Development

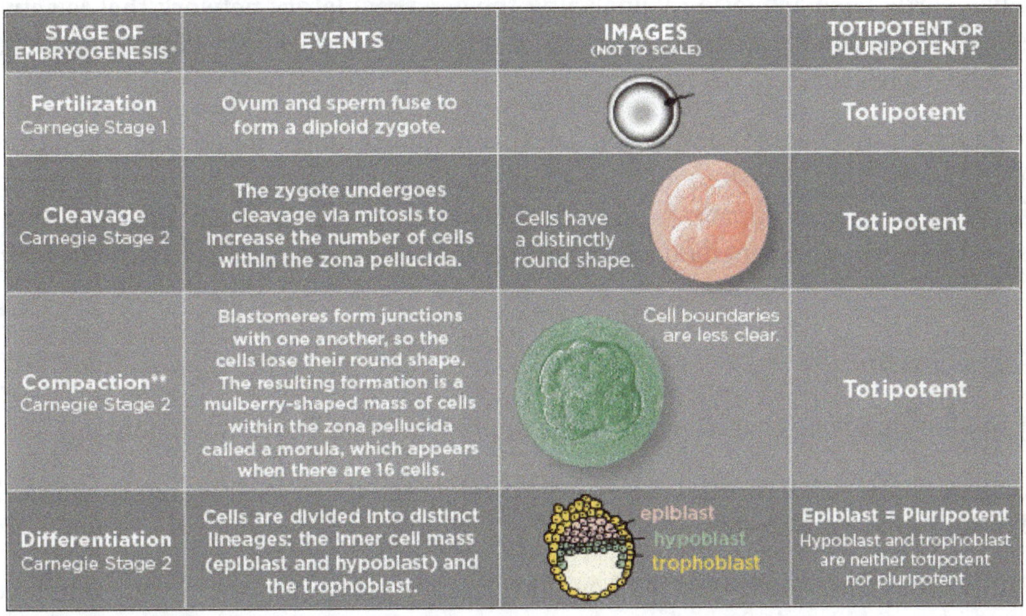

STAGE OF EMBRYOGENESIS*	EVENTS	IMAGES (NOT TO SCALE)	TOTIPOTENT OR PLURIPOTENT?
Fertilization Carnegie Stage 1	Ovum and sperm fuse to form a diploid zygote.		Totipotent
Cleavage Carnegie Stage 2	The zygote undergoes cleavage via mitosis to increase the number of cells within the zona pellucida.	Cells have a distinctly round shape.	Totipotent
Compaction** Carnegie Stage 2	Blastomeres form junctions with one another, so the cells lose their round shape. The resulting formation is a mulberry-shaped mass of cells within the zona pellucida called a morula, which appears when there are 16 cells.	Cell boundaries are less clear.	Totipotent
Differentiation Carnegie Stage 2	Cells are divided into distinct lineages: the inner cell mass (epiblast and hypoblast) and the trophoblast.	epiblast hypoblast trophoblast	Epiblast = Pluripotent Hypoblast and trophoblast are neither totipotent nor pluripotent

In figure, to provide a unified developmental chronology of mammalian embryology, the Carnegie Institution for Science developed a set of 23 stages known as the "Carnegie stages" whereby stages are defined by morphological development, and not directly by age or size. The number of cells at compaction varies between mammalian species.

When does a Totipotent Stem Cell Change?

Totipotent cells cease to exist around the time of the formation of the inner cell mass, when the trophectoderm lineage is established. Totipotent stem cells will divide and differentiate to give rise to cells that will develop into one of the three germ layers:

Cell	Potency	Fate
Epiblast	Pluripotent	The epiblast gives rise to the three primary germ layers (ectoderm, endoderm, and mesoderm) which will form all somatic lineages plus the germline.
Hypoblast	Not pluripotent, not totipotent	The hypoblast gives rise to the extraembryonic primitive endoderm, i.e. the yolk sac which provides nutrients to the embryo when primitive placental circulation has been established.
Trophoblast	Not pluripotent, not totipotent	The trophoblast gives rise to various extraembryonic structures which enable implantation into the uterine wall, secrete human chorionic gonadotropin to enable progesterone secretion from the corpus luteum, and form the chorion (fetal part of the placenta).

Features of Totipotent Stem Cells

Efforts have been made to establish methods to stabilize or create cells with an expanded developmental potential, relative to established lines of embryonic stem cells.

Through this work, typically in mice, researchers have found a few pieces of the totipotency puzzle. This has partly been achieved by analyzing the gene regulatory network that is active at the earliest stages of mammalian embryogenesis, and screening chemicals for their ability to modulate stem cell gene expression.

As it is impractical to assess totipotency in cells using the aforementioned gold standard in every instance (by transferring a blastomere to an empty zona pellucida, transferring it to a mother and seeing if it supports the development and birth of live young), some relevant measures and criteria have been established. These methods are based on characteristics specific to early developmental stages:

Gene Expression

Gene expression should be characteristic of the appropriate stage of development, e.g.:

- Genes specific to the two-cell stage have been identified in mice (Zscan4, Dux, Eifa, Zfp352, Tcstv1/3, and Tpodz1–5).
- Totipotent cells should lack the key pluripotency genes (Pou5f1, Sox2, and Nanog have been identified in mice).
- Oct4, a singular transcription factor, may be indicative of the developmental stage. Oct4 is encoded by POU5F1 and belongs to a family of transcription factors which activate the expression of their target genes. Oct4 is expressed in germ cells, embryonic stem cells and whole embryos, and expression levels vary dramatically during development. A key role for Oct4 in embryogenesis has been identified in several species, including humans.

Chromatin Mobility

Chromatin (the complexes of histones and DNA that condense to form chromosomes) mobility is higher in totipotent cells than in pluripotent cells, and chromatin dynamics seem to underlie changes in cellular plasticity.

Ability to Differentiate

Totipotent cells should have the ability to differentiate in vitro into cells representative of the three embryonic germ layers, as well as the trophectoderm, primitive endoderm and their derivatives.

Establishment of Expanded Potential Stem Cells

It has been noted that while some cells may pass several tests of totipotency, those which are "artificially" induced may not actually have a normal developmental counterpart. Therefore, a term has been coined to describe cells that retain features of totipotent stem cells: expanded potential stem cells (EPSCs).

Pluripotent Stem Cells

Pluripotent stem cells are master cells. They're able to make cells from all three basic body layers, so they can potentially produce any cell or tissue the body needs to repair itself. This "master" property is called pluripotency. Like all stem cells, pluripotent stem cells are also able to self-renew, meaning they can perpetually create more copies of themselves.

There are several types of pluripotent stem cells, including embryonic stem cells. At Children's Hospital Boston, we use the broader term because pluripotent stem cells can come from different sources, and each method creates a cell with slightly different properties. But all of them are able to differentiate, or mature, into the three primary groups of cells that form a human being:

- Ectoderm: Giving rise to the skin and nervous system.
- Endoderm: Forming the gastrointestinal and respiratory tracts, endocrine glands, liver, and pancreas.
- Mesoderm: Forming bone, cartilage, most of the circulatory system, muscles, connective tissue, and more.

Right now, it's not clear which type or types of pluripotent stem cells will ultimately be used to create cells for treatment, but all of them are valuable for research purposes, and each type has unique lessons to teach scientists. Scientists are just beginning to understand the subtle differences between the different kinds of pluripotent stem cells, and studying all of them offers the greatest chance of success in using them to help patients.

Types of Pluripotent Stem Cells

- Induced pluripotent cell (iPS cells).
- "True" embryonic stem cell (ES cells) derived from embryos.
- Embryonic stem cells made by somatic cell nuclear transfer (ntES cells).
- Embryonic stem cells from unfertilized eggs (parthenogenesis embryonic stem cells, or pES cells).

Mesenchymal Stem Cells

Bone marrow is a complex tissue containing hematopoietic cell progenitors and their progeny integrated within a connective-tissue network of mesenchymal-derived cells known as the stroma. The stroma cells, or Mesenchymal stem cells (MSCs), are multi-potent progenitor cells that constitute a minute proportion of the bone marrow, represented as a rare population of cells that makes up 0.001 to 0.01% of the total nucleated cells. They represent 10-fold less abundance than the haematopoietic stem cells, which contributes to the organization of the microenvironment supporting the differentiation of hematopoietic cells. MSC are present in tissues of young, as well as, adult individuals, including the adipose tissue, umbilical cord blood, amniotic fluid and even peripheral blood. MSCs were characterized over thirty years ago as fibroblast-like cells with adhesive properties in culture. The term MSC has become the predominant term in the literature since the early 90s, after which their research field has grown rapidly due to the promising therapeutic potential, resulting in an increased frequency of clinical trials in the new millennium at an explosive rate.

As data accumulated, there was a need to establish a consensus on the proper definition of the MSCs.

The International Society for Cellular Therapy has recommended the minimum criteria for defining multi-potent human MSCs. The criteria included: (i) cells being adherent to plastic under standard culture conditions; (ii) MSC being positive for the expression of CD105, CD73 and CD90 and negative for expression of the haematopoietic cell surface markers CD34, CD45, CD11a, CD19 or CD79a, CD14 or CD11b and histocompatibility locus antigen (HLA)-DR; (iii) under a specific stimulus, MSC differentiate into osteocytes, adipocytes and chondrocytes in vitro. These criteria presented properties to purify MSC and to enable their expansion by several-fold in-vitro, without losing their differentiation capacity. When plated at low density, MSCs form small colonies, called colony-forming units of fibroblasts (CFU-f), and which correspond to the progenitors that can differentiate into one of the mesenchymal cell lineages. It has been reported lately that MSCs are able to differentiate into both mesenchymal, as well as, non-mesenchymal cell lineages, such as adipocytes, osteoblasts, chondrocytes, tenocytes, skeletal myocytes, neurones and cells of the visceral mesoderm, both in vitro and in vivo.

All cells have half-lives and their natural expiration must be matched by their replacement; MSCs, by proliferating and differentiating, can be the proposed source of these new replacement cells as characterized in their differentiation capacity. This replacement hypothesis mimics the known sequence of events involved in the turnover and maintenance of blood cells that are formed from haematopoietic stem cells (HSCs). Unlike HSCs, MSCs can be culture-expanded ex vivo in up to 40 or 50 cell doublings without differentiation. A dramatic decrease in MSC per nucleated marrow cell can be observed when the results are grouped by decade, thus showing a 100-fold decrease from birth to old age. Being pericytes present in all vascularized tissues, the local availability of MSC decreases substantially as the vascular density decreases by one or two orders of magnitude with age. In recent years, the discovery of MSCs with properties similar, but not identical, to BM-MSCs has been demonstrated in the stromal fraction of the connective tissue from several organs, including adipose tissue, trabecular bone, derma, liver and muscle. It is important to note that the origin of MSCs might determine their fate and functional characteristics. Studies of human bone marrow have revealed that about one-third of the MSC clones are able to acquire phenotypes of pre-adipocytes, osteocytes and chondrocytes. This is in concordance with data showing that 30% of the clones from bone marrow have been found to exhibit a trilineage differentiation potential, whereas the remainder displays a bi-lineage (osteo-chondro) or uni-lineage (osteo) potential. Moreover, MSC populations derived from adipose tissue and derma present a heterogeneous differentiation potential; indeed, only 1.4% of single cells obtained from adipose-derived adult stem cell (ADAS) populations were tri-potent, the others being bi-potent or unipotent.

Effect of Mesenchymal Stem Cells on Immune cells

Mesenchymal Stem Cells have been shown to possess immunomodulatory characteristics, as described through the inhibition of T-cell proliferation in vitro. These observations have triggered a huge interest in the immunomodulatory effects of MSCs. The in vitro studies have been complemented in vivo, where both confirmed the immunosuppressive effect of MSC. MSC activating stimuli in vitro, appears to include the secretion of cytokines and the interaction with other immune cells in the presence of proinflammatory cytokines. Primarily, the in vivo effect has been originally shown in a baboon model, in which infusion of ex vivo–expanded matched donor or third-party MSCs delayed the time to rejection of histo-incompatible skin grafts. The delay indicated a potential role for MSC in the prevention and treatment of graft-versus-host disease (GVHD) in ASCT, in organ transplantation to prevent rejection, and in autoimmune disorders. Recently, MSCs were used to successfully treat a 9-year-old boy with severe treatment-resistant acute GVHD, further confirming the potent immunosuppressive effect in humans. The immunosuppressive potential has no immunologic restriction, whether the MSCs are autologous with the stimulatory or the responder lymphocytes or the MSCs are derived from a third party. The degree of MSC suppression is dose dependent, where high doses of MSC are inhibitory, whereas low doses enhance lymphocyte proliferation in MLCs. Broadly, MSC modulate cytokine production by the dendritic and T cell subsets DC/Th1 and DC/Th2, block the antigen presenting cell (APC) maturation and activation, and increase the proportion of $CD4^{+}CD25^{+}$ regulatory cells in a mixed lymphocyte reaction.

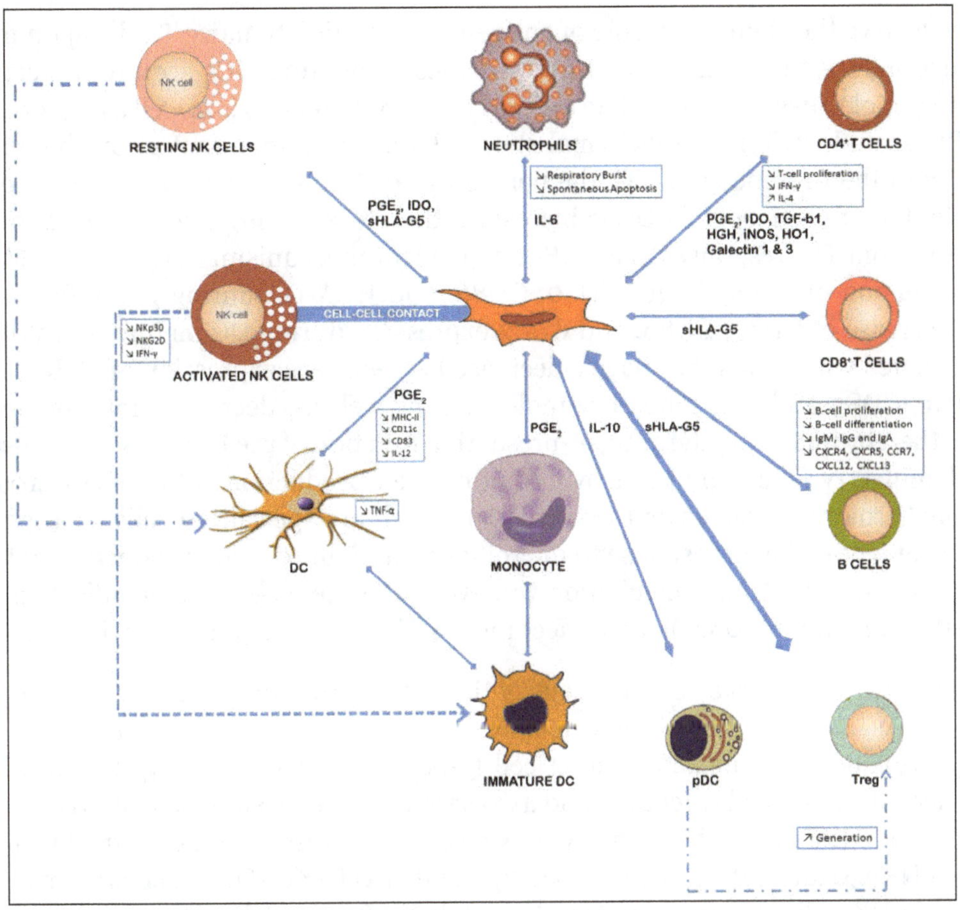

Figure shows potential mechanisms of the MSC interactions with immune cells. Mesenchymal stem cells (MSCs) can inhibit both the proliferation and cytotoxicity of resting natural killer (NK) cells, as well as, their cytokine production by releasing prostaglandin E2 (PGE2), indoleamine 2,3-dioxygenase (IDO) and soluble HLA-G5 (sHLA-G5). Killing of MSCs by cytokine-activated NK cells involves the engagement of cell-surface receptors (Thick blue line) expressed by NK cells with its ligands expressed on MSCs. MSCs inhibit the differentiation of monocytes to immature myeloid dendritic cells (DCs), bias mature DCs to an immature DC state, inhibit tumour-necrosis factor (TNF) production by DCs and increase interleukin-10 (IL-10) production by plasmacytoid DCs (pDCs). MSC-derived PGE2 is involved in all of these effects. Immature DCs are susceptible to activated NK cell-mediated lysis. MSC Direct inhibition of CD4+ T-cell function depends on their release of several soluble molecules, including PGE2, IDO, transforming growth factor-β1 (TGFβ1), hepatocyte growth factor (HGF), inducible nitric-oxide synthase (iNOS) and haem-oxygenase-1

(HO1). MSC inhibition of CD8+ T-cell cytotoxicity and the differentiation of regulatory T cells mediated directly by MSCs are related to the release of sHLA-G5 by MSCs. In addition, the upregulation of IL-10 production by pDCs results in the increased generation of regulatory T cells through an indirect mechanism. MSC-driven inhibition of B-cell function seems to depend on soluble factors and cell–cell contact. Finally, MSCs dampen the respiratory burst and delay the spontaneous apoptosis of neutrophils by constitutively releasing IL-6.

Immunomodulatory Effect of Mesenchymal Stem Cells in Innate Immunity

Dendritic cells have the elementary role of antigen presentation to naïve T cells upon maturation, which in turn induce the proinflammatory cytokines. Immature DCs acquire the expression of co-stimulatory molecules and upregulate expression of MHC-I and II, as well as, other cell-surface markers (CD11c and CD83). Mesenchymal stem cells have profound effect on the development of DC, where in the presence of MSC, the percentage of DC with conventional phenotype is reduced, while that of plasmacytoid DC is increased, therefore biasing the immune system toward Th2 and away from Th1 responses in a PGE-2 dependent mechanism. Furthermore, MSCs inhibit the up-regulation of CD1a, CD40, CD80, CD86, and HLA-DR during DC differentiation and prevent an increase of CD40, CD86, and CD83 expression during DC maturation. When mature DCs are incubated with MSCs they have a decreased cell-surface expression of MHC class II molecules, CD11c, CD83 and co-stimulatory molecules, as well as, decreased interleukin-12 (Il-12) production, thereby impairing the antigen-presenting function of the DCs. MSCs can also decrease the pro-inflammatory potential of DCs by inhibiting their production of tumour-necrosis factor α (TNF-α). Furthermore, plasmacytoid DCs (pDCs), which are specialized cells for the production of high levels of type-I IFN in response to microbial stimuli, upregulate production of the anti-inflammatory cytokine IL-10 after incubation with MSCs. These observations indicate a potent anti-inflammatory and immunoregulatory effect for MSC in vitro and potentially in vivo.

Natural killer (NK) cells are key effector cells of the innate immunity in anti-viral and anti-tumor immune responses through their Granzyme B mediated cytotoxicity and the production of pro-inflammatory cytokines. NK-mediated target cell lysis results from an antigen-ligand interaction realized by activating NK-cell receptors, and associated with reduced or absent MHC-I expression by the target cell. MSCs can inhibit the cytotoxic activity of resting NK cells by down-regulating expression of NKp30 and natural-killer group 2, member D (NKG2D), which are activating receptors involved in NK-cell activation and target-cell killing. Resting NK cells proliferate and acquire strong cytotoxic activity when cultured with IL-2 or IL-15, but when incubated with MSC in the presence of these cytokines, resting NK-cell, as well as, pre-activated NK cell proliferation and IFN-γ production are almost completely abrogated.

It is worth noting that although the susceptibility of NK cells to MSC mediated inhibition is potent, the pre-activated NK cells showed more resistance to the immunosuppressive effect of MSC compared to resting NK cells. The susceptibility of human MSCs to NK-cell-mediated cytotoxicity depends on the low level of cell-surface expression of MHC class I molecules by MSCs and the expression of several ligands that are recognized by activating NK-cell receptors. Autologous and allogeneic MSC were susceptible to lysis by NK cells, where NK cell-mediated lysis was inversely correlated with the expression of HLA class I on MSC. Incubation of MSCs with IFN-γ partially protected them from NK-cell-mediated cytotoxicity, through the up-regulation of expression of MHC-I molecules on MSCs. Taken together, a possible dynamic interaction between NK cells and

MSC in vivo exists, where the latter partially inhibit activated MSC, without compromising their ability to kill MSC, reflecting on an interaction tightly regulated by IFN-γ concentration.

Neutrophils play a major role in innate immunity during the course of bacterial infections, where they are activated to kill foreign infectious agents and accordingly undergo a respiratory burst. MSCs have been shown to dampen the respiratory burst and to delay the spontaneous apoptosis of resting and activated neutrophils through an IL-6-dependent mechanism. MSC had no effect on neutrophil phagocytosis, expression of adhesion molecules, and chemotaxis in response to IL-8, f-MLP, or C5a. Stimulation with bacterial endotoxin induces chemokine receptor expression and mobility of MSCs, which secrete large amounts of inflammatory cytokines and recruit neutrophils in an IL-8 and MIF-dependent manner. Recruited and activated neutrophils showed a prolonged lifespan, an increased expression of inflammatory chemokines, and an enhanced responsiveness toward subsequent challenge with LPS, which suggest a role for MSCs in the early phases of pathogen challenge, when classical immune cells have not been recruited yet. Furthermore, MSC have shown the capability to mediate the preservation of resting neutrophils, a phenomenon that might be important in those anatomical sites, where large numbers of mature and functional neutrophils are stored, such as the bone marrow and lungs.

Immunomodulatory Effect of Mesenchymal Stem Cells in Adaptive Immunity

T-cells are major players of the adaptive immune response. After T-cell receptor (TCR) engagement, T cells proliferate and undergo numerous effector functions, including cytokine release and, in the case of $CD8^+$ T cells (CTL), cytotoxicity. Abundant reports have shown that T-cell proliferation stimulated with polyclonal mitogens, allogeneic cells or specific antigen is inhibited by MSCs. The observation that MSCs can reduce T cell proliferation in vitro is mirrored by the in vivo finding through infusions of hMSCs that control GVHD following bone marrow transplantation. Nevertheless, there is no demonstrable correlation between the measured effects of MSCs in vitro and their counter effect in vivo due to the lack of universality of methodology correlating the in vitro findings with the in vivo therapeutic potential.

MSC inhibition of T-cell proliferation is not MHC restricted, since it can be mediated by both autologous and allogeneic MSCs and depends on the arrest of T-cells in the G0/G1 phase of the cell cycle. Thus, MSCs do not promote T-cell apoptosis, but instead maintain T cell survival upon subjection to overstimulation through the TCR and upon commitment to undergo CD95–CD95-ligand-dependent activation-induced cell death. MSC effects on T cell proliferation in vitro appear to have both contact-dependent and contact-independent components. Inhibition of T-cell proliferation by MSCs leads to decreased IFN-γ secretion in vitro and in vivo associated with increased IL-4 production by T helper 2 (TH2) cells. Taken together, there is an implication for a shift from a pro-inflammatory state characterized by IFN-γ secretion to an anti-inflammatory state characterized by IL-4 secretion. An imperative role for effector T-cell is the MHC restricted killing of virally-infected or of allogeneic cells mediated via $CD8^+$ CTLs, and which is down-regulated by MSC.

Regulatory T cells (Tregs), a subpopulation of T cells, are vital to keep the immune system in check, help avoid immune-mediated pathology and contain unrestricted expansion of effector T-cell populations, which results in maintaining homeostasis and tolerance to self-antigens. Tregs are currently identified by co-expression of CD4 and CD25, expression of the transcription factor FoxP3, production of regulatory cytokines IL-10 and TGF-β, and ability to suppress proliferation of activated $CD4^+CD25^+$ T cells in co-culture experiments. Differential expression of CD127 (α-chain of the IL-7 receptor) enable flow cytometry-based separation of human Tregs from CD127$^+$ non-regulatory T-cells. MSCs have been reported to induce the production of IL-10 by pDCs, which, in turn, trigger the generation of regulatory T cells. Furthermore, Tregs secrete TGF-β and when used in vitro, TGF-β in combination with IL-2 directs the differentiation of T-cells into Tregs, while Tregs suppress the proliferation of TCR-dependent proliferation of effector CD25^{null} or CD25^{low} T-cells in a non-autologous fashion. Also TGF-β alters angiogenesis following injury in experiments using MSC. In addition, after co-culture with antigen-specific T-cells, MSCs can directly induce the proliferation of regulatory T-cells through release of the immunomodulatory HLA-G isoform HLA-G5. Taken together, MSCs can modulate the intensity of an immune response by inhibiting antigen-specific T-cell proliferation and cytotoxicity and promoting the generation of regulatory T-cells.

Antibody producing B-cells constitute the second main cell type involved in adaptive immunity. Interactions between MSCs and B-cells have produced controversial results attributable to the inconsistent experimental conditions used. Initial reports on mice suggested that MSC exercise a dampening effect on the proliferation of B-cells, which is in concordance with most published works to date. Furthermore, MSCs can also inhibit B-cell differentiation and constitutive expression of chemokine receptors via the release of soluble factors and cell-cell contact mediated possibly by the Programmed Cell Death 1 (PD-1) and its ligand. The addition of MSCs, at the beginning of a mixed lymphocyte reaction (MLC), considerably inhibited immunoglobulin production in standard MLC, irrespective of the MSC dose employed, which suggests that third-party MSC are able to suppress allo-specific antibody production, consequently, overcoming a positive cross-match in sensitized transplant recipients.

However, other in vitro studies have shown that MSCs could support the survival, proliferation and differentiation to antibody-secreting cells of B-cells from normal individuals and from pediatric patients with systemic lupus erythematosus. A major mechanism of B-cell suppression was hMSC production of soluble factors, as indicated by transwell experiments, where hMSCs inhibited B-cell differentiation shown as significant impairment of IgM, IgG, and IgA production. CXCR4, CXCR5, and CCR7 B-cell expression, as well as chemotaxis to CXCL12, the CXCR4 ligand, and CXCL13, the CXCR5 ligand, were significantly down-regulated by hMSCs, suggesting that these cells affect chemotactic properties of B-cells. B-cell costimulatory molecule expression and cytokine production were unaffected by hMSCs. Regardless of the controversial in vitro effects, B-cell response is mainly a T-cell dependent mechanism, and thus its outcome is significantly influenced by the MSC-mediated inhibition of T-cell functions. More recently, Corcione et al have shown that systemic administration of MSCs to mice affected by experimental autoimmune encephalomyelitis (EAE), a prototypical disease mediated by self-reactive T cells, results in striking disease amelioration mediated by the induction of peripheral tolerance. In addition, it has been shown that tolerance induction by MSCs may occur also through the inhibition of dendritic-cell maturation and function, thus suggesting that activated T cells are not the only targets of MSCs.

Low concentrations of IFN-γ upregulate the expression of MHC-II molecules by MSCs, which indicates that they could act as antigen presenting cells (APCs) early in an immune response, when the level of IFN-γ are low. However, this process of MHC-II expression by MSCs, along with the potential APC characteristics, was reversed as IFN-γ concentrations increased. These observations could suggest that MSCs function as conditional APC in the early phase of an immune response, while later switch into an immunosuppressive function. Since bone marrow might be a site for the induction of T-cell responses to blood-borne antigens, and since MSC are derived from the stromal progenitor cells that reside in the bone marrow, therefore, MSC express a yet unidentified role in the control of the immune response physiology of the bone marrow. Dendritic cells are the main APC for T-cell responses, and MSC-mediated suppression of DC maturation would prohibit efficient antigen presentation and thus, the clonal expansion of T-cells. Direct interactions of MSCs with T-cells in vivo could lead to the arrest of T-cell proliferation, inhibition of CTL-mediated cytotoxicity and generation of $CD4^+$ regulatory T-cells. As a consequence, impaired $CD4^+$ T-cell activation would translate into defective T-cell help for B-cell proliferation and differentiation to antibody-secreting cells.

The hMSCs express few to none of the B7-1/B7-2 (CD80/CD86) costimulatory–type molecules; this appears to contribute, at least in part, to their immune privilege characteristic. Mechanisms that lead to immune tolerance rely on interrelated pathways that involve complex cross talk and cross regulation of T-cells and APCs by one another. Both soluble mediators and modulation exerted via complex networks of cytokines and costimulatory molecules appear to play a role in MSC regulation of T cells, and these mechanisms invoke both contact-dependent and -independent pathways.

Although many of the studies use MSC-conditioned medium, both contact-dependent and -independent mechanisms are probably invoked in the therapeutic use of MSCs. In addition to cytokines, the network of costimulatory molecules is hypothesized to play a prominent role in modulating tolerance and inflammation. MSCs down-regulate the expression of costimulatory molecules. The discovery of new functions for B7 family members, together with the identification of additional B7 and CD28 family members, is revealing new ways in which the B7:CD28 family may regulate T-cell activation and tolerance. Not only do CD80/86:CD28 interactions promote initial T-cell activation; they also regulate self-tolerance by supporting $CD4^+CD25^+$ Treg homeostasis. Cytotoxic T-lymphocyte antigen 4 (CTLA-4) can exert inhibitory effects in both B7-1/B7-2–dependent and –independent fashions. B7-1 and B7-2 can signal bi-directionally through engaging CD28 and CTLA-4 on T cells and by delivering signals into B7-expressing cells. The B7 family members—inducible co-stimulator (ICOS) ligand, PD-L1 (B7-H1), PD-L2 (B7-DC), B7-H3, and B7-H4 (B7x/B7-S1)—are expressed on professional APC cells, while B7-H4 and B7-H1 are expressed on hMSCs and on cells within non-lymphoid organs. These observations may provide a new means for regulating T-cell activation and tolerance in peripheral tissues. ICOS and PD-1 are expressed upon T-cell-induction, and they regulate previously activated T-cells. Both the ICOS:ICOSL pathway and the PD-1:PD-L1/PD-L2 pathway play a critical role in regulating T-cell activation and tolerance. There is consensus that both CTLA-4 and PD-1 inhibit T-cell and B-cell activation and may play a crucial role in peripheral tolerance. Both CTLA-4 and PD-1 functions are associated with Rheumatoid Arthritis (RA) and other autoimmune diseases. PD-1 is over expressed on $CD4^+$ T cells in the synovial fluid of RA patients. Whether or not these costimulatory molecules are critical mediators of MSC-mediated immune suppression and/or tolerance in vivo is still under current investigation.

Mesenchymal Stem Cells Escape the Immune System in Vitro

Studies have shown that MSCs escape the immune system, and this makes them a potential therapeutic tool for various transplantation procedures. MSCs express intermediate levels of HLA major histocompatibility complex (MHC) class I molecules, while they do not express HLA class II antigens of the cell surface. However, HLA class II is readily detectable by Western blot on whole-cell lysates of unstimulated adult MSCs, thus suggesting that MSCs contain intracellular deposits of HLA class II allo-antigens. Cell-surface expression can be induced by treatment of the cells with

IFN-γ for 1 or 2 days. Unlike adult MSCs, the fetal liver– derived cells have no intracellular nor cell surface HLA class II expression, but incubation with IFN-γ initiated their intracellular expression followed by surface expression. Differentiation of MSCs into their mesodermal lineages of bone, cartilage, or adipose tissue, both in adult and fetal MSCs continued to express HLA-I, but not class II. Undifferentiated MSCs in vitro fail to elicit a proliferative response from allogeneic lymphocytes, thus suggesting that the cells are not inherently immunogenic.

When pre-cultured with IFN-γ for full HLA class II expression, MSCs still escape recognition by allo-reactive T-cells, as is the case with MSCs differentiated adipocytes, osteoblasts, and chondrocytes. Limited in vivo data demonstrate the persistence of allogeneic MSCs into immunocompetent hosts after transplantation. In one patient treated with MSCs, DNA of donor MSC could not be detected in any organ at autopsy few weeks after the infusion, while in another patient receiving MSCs from two donors, the donor DNA from both donors was detected in lymph node and colon, the target organs of GVHD, within weeks after infusion. Data indicated that MSC were undetectable after two weeks in an allogeneic system.

Therefore, the question of whether MSCs are recognized by an intact allogeneic immune system in vivo remains open, although the in vitro data support the theory that MSCs escape the immune system. MSCs do not express FAS ligand or costimulatory molecules, such as B7-1, B7-2, CD40, or CD40L. When costimulation is inadequate, T-cell proliferation can be induced by the addition of exogenous costimulation. However, MSCs differ from other cell types, and no T-cell proliferation can be observed when they are cultured with HLA-mismatched lymphocytes in the presence of a CD28-stimulating antibody. However, in agreement with the in vitro data, infusion or implantation of allogeneic and MHC-mismatched MSCs into baboons has been well tolerated. Unique immunologic properties of MSCs were also suggested by the fact that engraftment of human MSCs occurred after intra-uterine transplantation into sheep, even when the transplantation was performed after the fetuses became immunocompetent. MSC mainly fail to activate T-cells and show to be targets for $CD8^+$ T cell-cytotoxicity, althought controversial. Phyto-hemagglutinin (PHA) blasts, generated to react against a specific donor, will lyse chromium-labeled mononuclear cells from that individual but it will not lyse MSCs derived from the same donor. Furthermore, killer cell inhibitory receptor (KIR ligand)–mismatched natural killer cells do not lyse MSCs. Thus, MSCs, although incompatible at the MHC, tend to escape the immune system.

Although MSCs are transplantable across allogeneic barriers, a delayed type hypersensitivity reaction can lead to rejection in xenogenic models of human MSCs injected into immunocompetent rats. In this study, MSCs were identified in the heart muscle of severe compromised immune deficiency rats, in contrast to that of immunocompetent rats. In the latter group, peripheral blood lymphocytes proliferated after re-stimulation with human MSCs in vitro, thus suggesting cellular immunization. Such a proliferative response in vitro has not been detected in humans treated with intravenous (IV) infusion of allogeneic MSCs.

Mechanisms of Immunosuppression by Mesenchymal Stem Cells

Several studies have acknowledged the immunosuppressive activities of MSCs, but the underlying mechanisms are far from being fully characterized. The initial step in the interaction between MSCs and their target cells involves cell–cell contact mediated by adhesion molecules, in concordance with studies showing the dependence of T-cell proliferation on the engagement of PD-1 by its ligand. Several soluble immunosuppressive factors, either produced constitutively by MSCs or released following cross-talk with target cells have been reported, including nitric oxide and indoleamine 2,3-dioxygenase (IDO), which are only released by MSC after IFN-γ stimulation with target cells, and thus not in a constitutive manner. IDO induces the depletion of tryptophan from the local environment, which is an essential amino acid for lymphocyte proliferation. MSC-derived IDO was reported as a requirement to inhibit the proliferation of IFN-γ-producing TH1 cells and together with prostaglandin E2 (PGE-2) to block NK-cell activity. In addition, IFN-γ, alone or in combination with TNF-α, IL-1α or IL-1β, stimulates the production of chemokines by mouse MSCs that attract T-cells and stimulate the production of inducible nitric-oxide synthase (iNOS), which in turn inhibits T-cell activation through the production of nitric oxide. It is worth noting that MSCs from IFN-γ receptor (IFN-γ-R1) deficient mice do not have immunosuppressive activity, which highlights the vital role of IFN-γ in the immunosuppressive function of MSC.

Additional soluble factors, such as transforming growth factor-β1 (TGF-β1), hepatocyte growth factor (HGF), IL-10, PGE-2, haem-oxygenase-1 (HO1), IL-6 and soluble HLA-G5, are constitutively produced by MSCs or secreted in response to cytokines released by target cells upon interacting with MSC. TNF-α and IFN-γ have been shown to stimulate the production of PGE-2 by MSCs above constitutive level. Furthermore, IL-6 was shown to dampen the respiratory burst and to delay the apoptosis of human neutrophils by inducing phosphorylation of the transcription factor signal transducer and activator of transcription 3, and to inhibit the differentiation of bone-marrow progenitor cells into DCs.

The failure to reverse suppression, when neutralizing antibodies against IL-10, TGF-β and IGF were added to MLR reactions does point to the possibility that MSC may secrete as yet uncharacterized immunosuppressive factors. Galectin-1 and Galectin-3, newly characterized lectins, are constitutively expressed and secreted by human bone marrow MSC. Inhibition of galectin-1 and galectin-3 gene expression with small interfering RNAs abrogated the suppressive effect of MSC on allogeneic T-cells. The restoration of T-cell proliferation in the presence of β- lactose indicates that the carbohydrate-recognition domain of galectins is responsible for the immunosuppression of T-cells and highlights an extracellular mechanism of action for the MSC-secreted galectins. In this respect, the inhibition of T-cell proliferation could result from either direct effects of galectin-1 and galectin-3 on T cells and/or through a direct or an indirect on effect on dendritic cells.

HLA-G5 represents another important molecule involved in MSC mediated regulation of the immune response, where its production has been shown to suppress T-cell proliferation, as well as NK-cell and T-cell cytotoxicity, while promoting the generation of Tregs. HLA-G protein expression is constitutive and the level is not modified upon stimulation by allogeneic lymphocytes in MSC/MLR. HLA-G5 is detected on MSCs by real-time reverse-phase polymerase chain reaction, immune-fluorescence, flow cytometry and enzyme-linked immunosorbent assay in the supernatant. Cell contact between MSCs and activated T-cells induces IL-10 production, which, in turn, stimulates the release of soluble HLA-G5 by MSCs. It is worth nothing that none of these molecules have an exclusive role and that MSC-mediated immune-regulation is a redundant system that is mediated by several molecules.

Mesenchymal Stem Cells in Response to Injury

One important characteristic of hMSCs is their ability to suppress inflammation resulting from injury, as well as, resulting from allogeneic solid organ transplants, and autoimmune disease. Consistent with in vitro studies, murine allogeneic MSCs are effective cellular therapy models in the treatment of murine models of human disease. Several studies have documented the substantial clinical improvements observed in animal models, when MSC were systemically introduced as a therapy in mouse models of multiple sclerosis, inflammatory bowel disease, infarct, stroke, and other neurologic diseases, as well as diabetes. These findings strongly suggest that xenogeneic hMSCs are not immunologically recognized by various immunocompetent mouse models of disease and are able to home to sites of inflammation. However, the mechanisms behind the immunosuppressive actions at the site of inflammation and its association with the homing activity have not yet been completely elucidated.

Nitric Oxide (NO) mediate its effect partly through phosphorylation of Stat-5, which results in suppression of T- cell proliferation, partly through the inhibition of NO synthase or the inhibition of prostaglandin synthesis. This reveals the MSC-dependent effects on proliferation. Although indoleamine-2, 3-dioxygenase (IDO) has been hypothesized to be critical in mediating the effect of NO, neutralizing IDO by using a blocking antibody did not interfere with NO's suppressive effects.

Within an in vivo setting, injury, inflammation, and/or foreign cells can lead to T-cell activation, which results in those T-cells producing proinflammatory cytokines including, but not limited to, TNF-α, IFN-γ, IL-1α, and IL-1β. Combinations of cytokines may also induce cell production of chemokines, some of which bind to CXCR3-R expressing cells (including T cells) that co-localize with MSCs. MSCs then produce NO, which inhibits Stat-5 phosphorylation, thereby leading to cell-cycle arrest (and thus halting T cell proliferation). In addition, iNOS appears to be important in mouse MSC in vivo effects. MSCs from mice that lack iNOS (or IFN-γR1) are unable to suppress GVHD. In contrast to mouse MSCs that use NO in mediating their immune-suppressive effect, hMSCs and MSCs from non-human primates appear to mediate their immune-suppressive effects via IDO. There is some controversy about whether the effect of IDO results from local depletion of tryptophan, or from the accumulation of tryptophan metabolites (which is suggested by data showing that use of a tryptophan antagonist, 1-methyl-L tryptophan, restored allo-reactivity that would otherwise have been suppressed by MSCs). In addition to its effect on the JAK-STAT pathway, NO may also influence mitogen activated protein kinase and nuclear factor κB, which would cause a reduction in the gene expression of proinflammatory cytokines.

Mesenchymal Stem Cell Clinical Applications

The clinical experience with and the safety of MSCs is of utmost interest for their wide therapeutic applications. The pioneering in vivo studies with MSC focused on the engraftment facilitation for the haematopoietic stem cells. Further work also focused on the regenerative functions of MSC in terms of functional repair of damaged tissues. Hypoimmunogenicity of MSC provided a critical advantage in their use for clinical and therapeutic purposes in vitro, followed by pre-clinical studies and reaching the human clinical studies with the use of allogeneic donors. Allogeneic MSC have proved to be an option with major advantages in clinical use, since the use of autologous MSC is hindered by the limited time frame for clonal expansion and the costly in vitro proliferation. However, some sub-acute conditions, such as autoimmune diseases, might allow the use of autologous MSCs and their culture in vitro. It is worth noting that some reports have recently challenged the belief that allogeneic MSCs are poorly immunogenic, indicating that in some cases an autologous MSC source could be advantageous. Recent reports have shown that MSCs from patients with autoimmune disease have a normal capability to support hematopoiesis, and to exercise immunomodulatory functions, and to show normal phenotypic characteristics.

The perspective role of adult stem cells in degenerative disease conditions, where there is progressive tissue damage and an inability to repair, possibly due to the depletion of stem cell populations or functional alteration, has been considered. In cases of osteoarthritis, a disease of the joints where there is progressive and irreversible loss of cartilage characterized by changes in the underlying bone, Murphy et al showed that the proliferative capacity of the MSC was substantially reduced, and this was independent of the harvest site from patients with end-stage OA undergoing joint replacement surgery. In this study the marrow samples were harvested both from the site of surgery (either the hip or the knee) and also from the iliac crest. These effects were apparently disease-related, and not age-related. However, the data suggest that susceptibility to OA and perhaps other degenerative diseases may be due to the reduced mobilization or proliferation of stem cells. In addition, successfully recruited cells may have a limited capacity to differentiate, leading to defective tissue repair. Alternatively, the altered stem cell activity may be in response to the elevated levels of inflammatory cytokines seen in OA, which was confirmed by several other investigators.

Similarly, the functional impairment of the anti-proliferative effect of MSCs derived from patients with aplastic anaemia or multiple myeloma might be resulting from an intrinsic abnormality in the microenvironment of the bone marrow, which is consistent with the possible use of autologous MSC for therapeutic purposes.

With the knowledge of the homing capacity of MSC and their capacity to engraft into the recipient's bone after systemic administration, MSCs have been utilized to treat children with severe osteogenesis imperfecta, leading to improved parameters of increased growth velocity and total body mineral content associated with fewer fractures. Systemic infusion of allogeneic MSCs also led to encouraging bone marrow recovery in patients with tumors following chemotherapy. The immunosuppressive effect of infused MSCs has been successfully shown in acute, severe graft-versus-host disease (GvHD). The probable effect of MSC was the inhibition of donor T-cell reactivity to histocompatibility antigens of the recipient tissue. Currently, there is no successful therapy for steroid-refractory acute GVHD. The possible role of MSCs in this context is therefore of potential interest. Le Blanc et al reported a case of grade IV acute GVHD of the gut and liver in a patient who had undergone ASCT with cells from an unrelated female donor. The patient was unresponsive to all types of immunosuppression drugs. When the patient was infused with 2x 10^6 MSCs per kilogram from his HLA-haploidentical mother, his GVHD responded with a decline in bilirubin and normalization of stools. After the MSC infusion, DNA analysis of his bone marrow showed the presence of minimal residual disease. When immunosuppression was discontinued, the patient again developed severe acute GVHD, with its associated symptoms within a few weeks.

Modulation of host allo-reactivity led to accelerated bone-marrow recovery in patients co-transplanted with MSCs and haplo-identical HSCs. Clinical trials are being conducted on the immunomodulatory potential of MSCs in the treatment of Crohn's disease, with the potential for those cells to contribute to the regeneration of gastrointestinal epithelial cells.

MSCs are characterized by their hypoimmunogenicity. In 2000, data from several research groups demonstrated long-term allo-MSC engraftment in a variety of non-cardiac tissues in the absence of immunosuppression. On the basis of these observations, investigators began to look into the possibility of allo-MSCs engraftment into affected myocardium in rats, and later in swine, where allo-MSCs were found to readily engraft in necrotic myocardium and favorably alter ventricular function. The allo-MSC engraftment occurred without evidence of immunologic rejection or lymphocytic infiltration in the absence of assisted immunosuppressive therapy emphasizing some of the apparent advantages of these cells over other cell populations for cellular cardiomyoplasty. The immunologically privileged status of MSCs was also observed in xenogeneic setting, where Saito et al injected MSC intravenously from C57BL/6 mice into immunocompetent adult Lewis rats. When these animals were later subjected to MIs, murine MSCs could be identified in the region of necrosis, and these cells expressed muscle specific proteins not present before coronary ligation.

Animal Models

Consistent with results from in vitro studies, murine allogeneic MSCs are effective in the treatment of murine models of human disease. Several studies have reported clinical improvements in mouse models of multiple sclerosis and amyotrophic lateral sclerosis, inflammatory bowel disease, stroke, diabetes, infarct and GVHD using I.V injected xenogeneic hMSCs rather than allogeneic MSCs. A major advantage in using hMSCs in mouse models of human disease is that the possibility of gathering mechanistic data through measuring biomarkers from body fluids or using noninvasive imaging technology, which may prove to be an advantage in clinical studies when applied on humans.

In experiments designed to study the trafficking of hMSCs, investigators used mouse models of severe erosive polyarthritis characterized by an altered transgene allele that results in chronic over-expression of TNF-α and which resemble human RA patients. The motive behind utilizing these mice models was to investigate similarities in MSC homing with mouse models of chronic asthma and acute lung injury. Injected hMSC revealed a reduction in ankle arthritis parameters associated with decrease appendage related erythema, possibly due to the MSC localization to ankle joints as revealed by bioluminescence. Similar observations for inducing tolerance were made using adipose-derived MSC, where Treg were induced in RA PBMC and in mouse models of arthritis. Furthermore, studies of rheumatoid arthritis T-cells showed a down-regulation of effector response using adipose-derived MSCs. Variations in this potential described by the capability of MSCs to down-regulate collagen-induced arthritis, and in the ability to induce Tregs, depend on the source of MSC (mouse vs. human) and its characteristics (primary isolate of MSC line), which reflect on difference in function compared to primary expanded MSC. Other studies reported that in the collagen-induced model of arthritis, mice infused with MSCs have increased numbers of $CD4^{+}CD25^{+}$ cells that express FoxP3 and thus reveal a Treg phenotype. Recent data on collagen-induced arthritis model, where murine MSCs did not reveal therapeutic benefits against arthritis in vivo, but did show anti-proliferative effect in vitro suggest that there is no appropriate in vitro measures that can be accurately extrapolated into a potential therapeutic utility of MSCs in vivo, and that mouse MSCs show difference in functional characteristics to hMSC in terms of immunoregulatory capacity.

MSC's immunological properties appeared to have potential therapeutic advantages in other forms of autoimmune diseases, especially in type 1 diabetes. In NOD mouse model, several physiological defects that aim to maintain peripheral and central tolerance contribute to the development of autoimmune diabetes. These defects are summed up as a combination of immune cell dysfunction (including T-cell, NK cells, B-cells, and dendritic cells), associated with the presence of inflammatory cytokine milieu. MSCs possess specific immunomodulatory properties capable of halting autoimmunity through immunomodulation processes. The processes might be through a direct effect via the presentation of differential levels of negative costimulatory molecules and the secretion of regulatory cytokines that affect regulatory T-cells/autoreactive T-cells. Furthermore, MSCs could modulate the immunological dysregulation observed in antibody producing B-cells and cytotoxic NK cells. Dendritic cells have been shown to be defective in NOD mice characterized by higher levels of costimulation with a potential capability to shift to a TH-1 type of immune response.

In an experimental mouse model of diabetes induced by streptozotocin, it was observed that MSCs promoted the endogenous repair of pancreatic islets and renal glomeruli. Similarly, co-infusion of MSCs and bone-marrow cells inhibited the proliferation of β-cell-specific T-cells isolated from the pancreas of diabetic mice and restored insulin and glucose levels through the induction of recipient-derived pancreatic β-cell regeneration in the absence of trans-differentiation of MSCs. These studies show that the in vivo administration of MSCs is clinically efficacious through the modulation of pathogenic β- and T-cell responses and through potent bystander effects on the target tissue.

The timing of MSC infusion seems to be a critical parameter in their therapeutic efficacy. In the EAE mouse model of multiple sclerosis, MSC systematically injected at disease onset ameliorated myelin oligodendrocyte glycoprotein (MOG)-induced EAE and further decreased the infiltration T-cells, B-cells and macrophages into the central nervous system (CNS). Furthermore, T cells isolated from the lymph nodes of MSC-treated mice did not proliferate after in vitro re-challenge with MOG peptide, which is an indication of the induction of T-cell anergy. Systematic injection of MSCs was found to inhibit the in vivo production of pathogenic plp-specific antibodies and to suppress the encephalitogenic potential of plp-specific T cells in passive-transfer experiments. In this model, the MSCs migrated to the lymphoid organs, as well as, to the inflamed CNS, where they exercised a protective effect on the neuronal axons in situ. In these studies, the therapeutic effect of MSCs depended on the release of anti-apoptotic, anti-inflammatory and trophic molecules, as occurred in the case of stroke in rats, and, possibly, on the recruitment of local progenitors and their subsequent induction to differentiate into neural cells. As trophic effect, the MSCs appeared to favor oligo-dendrogenisis by neural precursor cells.

Several other studies have provided insights into the effects of MSCs mediated by cytokines. In a model of acute renal failure, the administration of MSCs increased the recovery of renal function through the inhibition of production of proinflammatory cytokines, such as Il-1β, TNF and IFNγ, and through an anti-apoptotic effect on target cells. Along the same line, the anti-inflammatory activity of MSCs was revealed in a mouse model of lung fibrosis, where they inhibited the effects of IL-1α-producing T cells and TNF-producing macrophages through the release of IL-1 receptor antagonist (IL-1RA). The release of trophic factors such as the WNT-associated molecule secreted frizzled-related protein 2 (SFRp2), which leads to the rescue of ischemic cardiomyocytes and the restoration of ventricular functions represent another important function for MSC.

With all the promising therapeutic potential of MSC, there seems to be a growing concern about their association with tumors. The immunoregulatory and anti-proliferative effects of MSCs led to several studies investigating the inhibitory effect of MSCs on tumor growth. Inhibition or, more frequently, stimulation of tumor-cell proliferation in vitro and/or tumor growth in vivo by MSCs has been reported. The heterogeneous nature of the MSC populations and the different experimental tumor models used, contribute to the effect of tumors on MSC in which the microenvironment generated by tumors influence the behavior of MSCs. Two main mechanisms are probably involved in the enhancement of tumor growth by MSCs.

First, the cell-to-cell cross-talk between MSCs and tumor cells contribute to tumor progression, thus integrating within the tumor stroma, and second, the suppressive effects of MSCs on the immune system of tumor-bearing hosts might facilitate tumorigenesis, as shown for the inhibition of melanoma rejection, possibly mediated by regulatory $CD8^+$ T cells. Irrespective of the possible interactions between cancer cells, immune cells and MSCs, the potential risk of stimulating the growth cancer by MSCs must be considered.

Identification and Isolation of Stem Cells

Every type of stem cell or progenitors expresses specialized cell surface receptors and intracellular proteins. These receptors or markers have the capability to selectively bind to other protein and ligand which in turn starts a particular signalling pathway. These cell surface receptor are stem cell markers and are known as cluster of differentiation (CD). Each type of cell has a specific combination of receptor or markers which makes them distinguishable from all other types of cells. We can use this uniqueness of stem cells for identification, characterization and isolation. Markers or receptor for stem cells are dependent on the origin of the stem cells and are very specific. For e.g., hematopoietic stem cells are CD34+, Sca-1+, c-kit+, and CD43+ while skeletal muscle stem cells are CD34+/-, Sca-1+, c-kit-, and CD43-.

Isolation of Stem Cells

To isolate these rare cell populations of stem cells, techniques based on cell surface markers cell can be utilized. One such technique is magnetic activated cell sorting (MACS) which is used for the separation of cells based on specific cell surface markers. Cell isolated form a tissue are generally separated on the basis of density gradient centrifugation using Histopaque or Ficoll to remove debris or unwanted cell population. To enrich desired cell population, Cell mixture gets stained with magnetic beads attached to cell surface specific antibodies is separated with the help of high gradient magnetic columns. Cells can be selected and sorted based one or two surface marker with MACS. Fluorescence activated cell sorting (FACS) is used for the isolation, sorting and analysis of various population of cells based on the different combination of cell surface marker. FACS is based on light scattering properties of cells and fluorescent labelled cell surface markers.

Stem Cell Use

Stem cells can be utilized as basic tools by scientists and researchers to help understand and resolve several problems in the fields of biology and medicine. Different kind of stem cells either fetal, cord blood, embryonic or adult can be used in research and to provide us information and understanding of human and animal development and disease pathogenesis, can help in developing novel therapies for different disorders and to screen novel therapeutic compounds. Bone marrow transplantation is a very well-established treatment for variety of blood disorders like leukemia which involves the transplantation of bone and blood stem cells. Stem cells can be used to generate healthy and functioning specialized cells, which can then replace diseased or dysfunctional cells. It is similar to the process of organ transplantation only the treatment consists of transplanting cells instead of organs.

Stem Cell Therapy

Cancers and Leukemia

Hematopoietic stem cell transplantation (HSCT) is the procedure of intravenous infusion of autologous or allogeneic stem cells to re-establish haematopoiesis in patients having damaged bone marrow or defective immune system. The patient receives stem cells from his/her own blood in autologous stem cell transplant and in case of allogeneic stem cell transplant, the patient receives stem cells from a donor (related or unrelated). Syngenic stem cell transplant relies on the patient receiving stem cells from his/her identical twin.

When a patient receives an organ from a donor during transplant surgery, it is essential to get a HLA typing done. Human leukocyte antigen (HLA) is genetically determined cell surface molecule, which is responsible for the development of immunological response against the foreign antigens. The patient's immune system detects the antigens on the cells of the organ if they are different or not "matched" resulting in transplant rejection. The more similar the antigens are between the donor and recipient, the less likely that the organ will be rejected.

The first successful transplantation was done by Dr. E. Donnall Thomas in 1959 viz. syngeneic bone marrow transplantation in two patients with acute lymphoblastic leukemia. India's first successful allogeneic BMT was done on 20th March 1983 at Tata Memorial Hospital on a nine year old girl having acute myeloid leukemia. More than 50,000 HSCTs— 53% autologous and 47% allogeneic—are performed every year worldwide, according to the World Wide Network of Blood and Marrow Transplantation.

Depending upon the stem cell source, HSCT can be sub-divided into Bone Marrow Transplantation (BMT), Peripheral Blood Stem Cell Transplantation (PBSCT) and Umbilical Cord Stem Cell Transplantation (UCSCT). Bone marrow can be considered the best source for the isolation of HSCs. The donor is punctured in the bone (typically hipbone) and blood is drawn out of the bone marrow with the help of a syringe. In practice, around 3×10^8 nucleated cells/kg of the recipient's body weight are being harvested from the bone marrow. Bone marrow stem cells account for around 50% of paediatric transplants and less than 15% of adult transplants. Yet, the isolation of stem cells is a painful and tedious process from the bone marrow, especially with young donors. Apart from bone marrow, peripheral blood (mobilized) is also used. HSCs can be isolated from the peripheral blood by leukapheresis after stem cell mobilization using GM-CSF (Granulocyte Monocyte- Colony Stimulating Factor) in a dose of 5µg/kg/day (divided into two doses for 4-5 days). 5×10^8 mononuclear cells /kg or for CD34+ CELLS $2\text{-}5 \times 10^6$ are harvested. Currently around 70% PBSCs of related donor and 60% of unrelated donor are used in HSCT. Nowadays, umbilical cord blood is emerging on the forefront of HSCs isolation. UCB stem cells can be harvested during delivery time by clamping the umbilical vein. Normal blood bags can be used in the collection of UCB, which are then cryopreserved in liquid nitrogen for further use.

Heart Diseases

Cardiovascular disorders are a big cause of death worldwide which claims more than 17 million deaths per year. Acute Myocardial infarction or heart attack happens if the blood supply to heart gets blocked often due to blood clots and the therapies like angioplasty and insertion of stent do not fully solve the issue. Researchers and scientists are trying to cure heart diseases with the help of stem cells which can repair and replace the damaged tissues causes by heart failure and heart attacks. It has been shown and proven that stem cell therapy for the treatment of various heart diseases is safe. Several clinical trials has been done to establish stem cell treatment for heart ailments and has been successfully proven its efficacy. It has been found that stem cells has the potential to get converted into cardiomyocytes which in turn helps in the regeneration of heart muscles. Adult bone marrow hematopoietic stem cells once transplanted into the damaged walls of heart ventricles leads to the formation of new cardiomyocytes, smooth muscle cells and vascular endothelium which ultimately helps in generating de novo myocardium that includes coronary arteries, arterioles, and capillaries. Embryonic stem cells and induced pluripotent stem cell can also be used for the treatment of heart diseases.

Neurological Diseases

Parkinson's disease, Huntington's disease, stroke, amyotrophic lateral sclerosis are few of the neurodegerative diseases which are characterized by changes like apoptosis in neuron that are involved in networking of brain and signals. These changes can leads to paralysis and sensation loss around the site of injury. Stem Cells can be safe and effective treatment strategy for these disorders where conventional treatments are not very successful. Stem Cell therapy is becoming a boon for the treatment of various neurological disorders like Muscular Dystrophy, Brain stroke, Autism, Cerebral Palsy Dementia, Head Injuries, Neurological infections, Motor neuron diseases, Genetic neurological disorders. Research conducted on rats having stroke suggested that if specific drugs were administration which increases stem cell division differentiation recovery is fast. Scientist and clinicians also treated the spinal cord injury patient by injected the stem cells isolated from adult bone marrow and umbilical cord blood into damaged areas of the spinal cord. Mesenchymal stem cell transplant in spinal cord shown that there is delay motor neuron decline and improvement in the performance of motor neuron in mouse models of Amyotrophic Lateral Sclerosis, a type of neurodegerative diseases. Stem cell injection in cerebrospinal fluid of multiple sclerosis models decreased clinical symptoms and reversed disability.

Type I Diabetes

Diabetes is a destructive and composite metabolic disease which is expected to affect more than 500 million people globally by the year 2030. People with Diabetes type 1 has lack of insulin while in Type 1 diabetes, an autoimmune disorder, cells of the pancreas that generally manufacture insulin gets damaged by patient's own immune system. It mostly happens in kids and also known as juvenile onset or insulin dependent diabetes. In absence of insulin, high blood sugar causes symptoms like thirst, fatigue and hunger, which in turn can have consequences, like damage to the nerves, blood vessels and internal organs. Lack of donors makes it difficult to transplant insulin-producing pancreatic β-cells. Scientific research studies show that it may be possible differentiates human embryonic and adult stem cells into insulin-producing cells that ultimately can be used for transplantation in diabetes patients to treat diabetics. Transplanted stem cells to treat diabetes I works as a two way therapy. First, they have the potential to regenerate insulin producing beta cells and helps in the secretion of insulin. Second, these cells can modulate the immune system which is comprised here by inhibitions of responses that leads to autoimmune attack on the pancreatic cells.

Liver Diseases

Chronic liver disease is the fifth largest killer all over the world. For a variety of liver diseases like liver cirrhosis, liver failure or cancer, liver transplantation is the only effective treatment exists till date. But the availability of donor for liver transplant is a very big challenge in this area. Stem cells transplantation holds assurance as a substitute to treat liver disorders. Scientists proved the hepatocytes can be grown outside body from stem cells by transdifferetiation. Adipose tissues and bone marrow from the patients itself are a very good source of stem cells that can be used in the treatment of liver diseases. Transplanted stem cells regenerate and repair scarred areas of liver and also produce required growth factors for better repair. According to various studies transplanted cells improves the liver conditions that is depicted by normalized levels of various liver enzymes and ultrasound.

Rheumatoid Arthritis

Rheumatoid Arthritis is a type of autoimmune disorders where joint destruction happens accompanied by other complications like pulmonary fibrosis, renal damage etc. Current treatments provide temporary relief from pain but also have an adverse effect due to inhibitions of immune response. Stem cell therapy has been demonstrated as an effective therapy which induces healing in various types of arthritis. Mesenchymal stem cells, a type of stem cells were isolated bone marrow home to the inflamed and injured tissue and starts producing anti-inflammatory agents thus help in healing and regeneration of tissues.

Baldness

It has already been proven that our hair follicles are a source of stem cells. This concept is very helpful in the treatment of baldness which is major concern of people now days. Dermatologists and scientist already shown that stem cell therapy is a very good approach for the treatment of loss of the hair and stem cells replenishments which has been lost. Dermatologists have found that stem cells transplant into the pores of the hair scalp, skin regeneration on the scalp happens that send molecular signals to the follicle that results in growth of new hairs. Stem cells for this treatment can be taken from bone marrow, adipose tissue or from follicles of existing hairs.

Corneal Diseases and Blindness

Blindness or various other eye disorders causes loss of visual function. These can be caused by failure of the retina to capture light or failure of the light path to reach the retina signal transmission to the brain. Various factors like age related macular degeneration (ARMD), cataracts, diabetic retinopathy and glaucoma can cause blindness. Loss of vision may have a vast impact on anyone's life, but very limited treatments are available for these disorders. Retinal stem cell are coming up in the limelight now a days as a treatment option which give a hope to restore vision. Scientific studies showed that embryonic, adult stem cells taken from patient, a relative or cadaver has the ability to grow in laboratory and after transplantation over the damaged area, stimulate the neural repair ultimately improving vision of the patient.

Dental Stem Cells for Missing Teeth

Teeth are very important part of the human body and are composed of three types of highly mineralized hard tissues: enamel, dentin and cementum. Teeth are formed by dental epithelium and cranial neural crest mesenchymal cells with communications in between. Tooth loss is a main challenge in contemporary dentistry. Bioengineering of teeth is an important goal of regenerative medicine. At present, missing teeth are restored by dentures or dental implants prepared from synthetic materials. Although these prostheses serve the function, it may cause denture-induced stomatitis and traumatic ulcers. To overcome these limitations, the new advance of tooth regeneration or engineering, which are based on stem cell biology, has been recommended by researchers and clinicians. Tooth engineering by stem cells is through isolation and culture under in vitro conditions and differentiation into particular cells that will from dentin, enamel and will perform tooth morphogenesis. Stem cells isolated from the pulp of human exfoliated deciduous teeth have the capacity to induce bone formation, generate dentine and differentiate into other non-dental mesenchymal cell derivatives in vitro.

Stem Cells for Drug Delivery and Gene Therapy

Although this area of stem cell biology has not been much explored but stem cells holds a great promise in the drug delivery and gene therapy. Stem cells isolated from patients can be genetically manipulated with therapeutic gene under in vitro conditions. Upon transplantation stem cells will differentiated into the desired cell type and the modifies gene will be active. According to one scientific research, skin cells from an immunodeficient mouse generated cellular therapy that partially restored normal function in the mouse.

Induced Pluripotent Stem Cells (iPSC)

Induced pluripotent stem cells are actually the adult cells which has been genetically reprogrammed into an embryonic stem cell like state and also express genes proteins which are essential for maintaining properties like of embryonic stem cells.Takahashi and Yamanaka reported that pluripotent stem cells can be established by 4 transcription factors which they selected from a list of 24 mouse embryonic stem cells enriched genes: Oct4, Sox2, Klf4 and Myc. Various combinations of these and other transcription factors has been used after this by other scientists. Direct reprogramming generated induced pluripotent stem cells that were phenotypically functionally equal to embryonic stem cells. Mouse iPSCs demonstrate important characteristics of pluripotent stem cells, including expressing stem cell markers, forming tumors containing cells from all three germ layers, and being able to contribute to many different tissues when injected into mouse embryos at a very early stage in development. Human iPSCs also express stem cell markers and are capable of generating cells characteristic of all three germ layers. iPSC has the ability to differentiate into many cell types and are easily accessible. These types of cells are Individual or patient specific that means these are personalized or non-immunogenic. These cells can be useful for understanding mechanisms of disease and will help in drug, toxicity testing.

Cancer Stem Cells

Cancer arises from the unrestrained expansion and division of cells which may be caused by the teamwork of mutations in genetic materials that may other activate genes that leads to uncontrolled cell division. An anti-cancer therapy tries to inhibit cancer cell growth, causing cancer cells death. Commonly, an initial cancer treatment kills the cancer cells and patients improve but chances of relapse of the disease are always there. Within any tumour population, many different kinds of cancer cell exist, only a few gets targeted by a particular treatment. The cancer stem cell concept suggests that tumour growth is driven by a rare subpopulation or subset of cells, known as cancer stem cells. These cancer stem cells do not divide rapidly, hence are not targeted by chemotherapy. Thus cancer stem cells survives conventional chemotherapy and divides rapidly to form a new tumor in this way there is relapse and re growth of tum ors. Better understanding the nature and biology of cancer stem cells will leads to the identification of molecular targets against these cells that will be important for future cancer therapies.

5

A Comprehensive Study of Genetic Engineering

Genetic engineering is a set of technologies used to modify the genome of organisms. It includes the transfer of genes from one organism to another within or across species. This chapter discusses in detail the major concepts of genetic engineering such as gene expression, gene cloning, recombinant proteins, and cell and tissue engineering.

Genetic Engineering

All the living organisms may be classified in to two broad categories- Prokaryotes and eukary-otes. The cell has unique characteristics of genetic material duplication and cell division, thus increasing the number of cells. All these processes are precisely regulated at gene level. A gene is basic unit of inheritance, made up of deoxyribonucleic acid and constitutes part of chro-mosome. Genes are responsible for production of traits/phenotypes by production of protein via a tightly regulated process of transcription and translation. The science which deals with study of genes and inheritance is genetics. Genetics has contributed most to the development of biotechnology. Ample food, better healthcare facilities, production of commercial substanc-es, clean ecosystem and low cost energy are some of the area where genetics have played sig-nificant role. Application of scientific and engineering principles to biological phenomena for commercial production of goods and services have led to development of Biotechnology. In the last 40 years, astonishing progress has been made in the field of biotechnology and most of this progress is attributed to genetic engineering. Genetic engineering is the manipulation of genetic material or the product of genes for the welfare of humans and living beings as well as ecosystem.

The word 'Genetic Engineering' was coined in the science fiction novel 'Dragon's Island' by Jack Williamson in 1951. Although it was Peter Lobban who first proposed but the tech-nique was first demonstrated by Cohen and Boyer in 1972 in the research paper 'Biochemical method for inserting new genetic information into DNA of Simian Virus 40: Circular SV40 DNA molecules containing Lambda Phage genes and the galactose operon of E. coli' and 'Con-struction of biologically functional bacterial plasmids in vitro' in 1974. With the discovery, isolation and application of restriction endonucelases by Arber, Nathans and Smith, recombi-nant DNA technology was made possible and this discovery honoured them with Nobel Prize in medicine in 1978. The first patent for production of biologically functional chimeric DNA was granted to Cohen and Boyer in 1980. This was immediately followed by production of re-combinant human insulin Eli Lilly in 1982 which transformed the way diabetic patients were treated.

Recombinant DNA technology made it possible to make millions of copies of specific DNA product (molecular cloning) and also production of recombinant protein in suitable host. Therefore, this technology is lauded as most revolutionary technology of twentieth century. There are two type of genetic material i.e. DNA and RNA depending upon the type of organism. All the activities of a cell or a multicellular body are dependent on the activity of gene which is the conversion of information from the DNA to the actionable form protein via an intermediate RNA. This flow of information is known as "central dogma of life".

Central Dogma of Life

"Central dogma is the process in which the genetic information flows from DNA to RNA, to make a functional product protein."

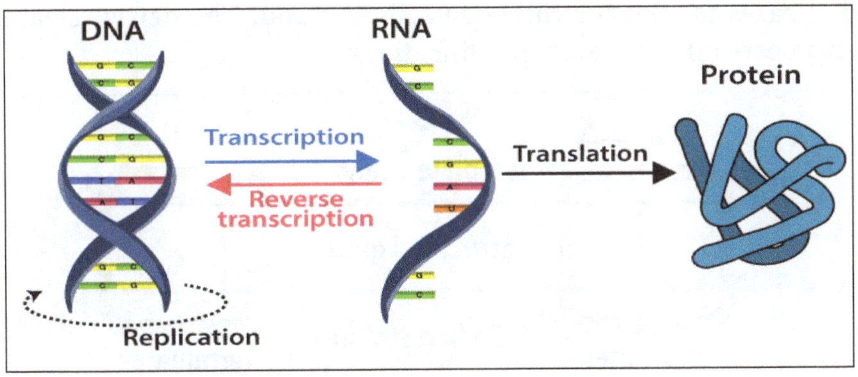

The central dogma illustrates the flow of genetic information in cells, the DNA replication, and coding for the RNA through the transcription process and further RNA codes for the proteins by translation.

The concept of a sequence of interaction can be understood through the framework. The most common includes biopolymers. The major category of biopolymers include Proteins, RNA and DNA that are further divided into general transfers, unknown transfers, and special transfers. Special transfers occur in an exceptional case in the laboratory. General transfer occurs in almost all cells. It describes the regular flow of information through transcription and translation. Unknown transfers are said never to occur.

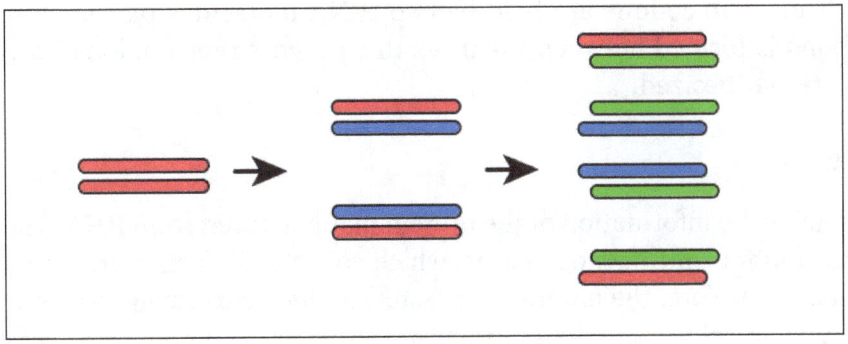

The new DNA strands are formed, with one strand of the parent DNA and the other is newly synthesized, this process is called semiconservative DNA replication.

Central Dogma Steps

The central dogma takes place in two different steps:

Transcription

Transcription is the process by which the information is transferred from one strand of the DNA to RNA by the enzyme RNA Polymerase. The DNA strand which undergoes this process consists of three parts namely promoter, structural gene, and a terminator.

The DNA strand that synthesizes the RNA is called the template strand and the other strand is called the coding strand. The DNA-dependent RNA polymerase binds to the promoter and catalyzes the polymerization in the 3′ to 5′ direction. As it approaches the terminator sequence, it terminates and releases the newly synthesized RNA strand. The newly released RNA strand further undergoes post-transcriptional modifications.

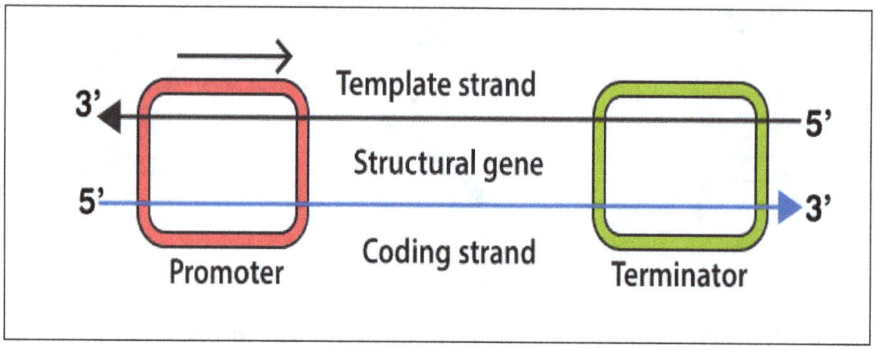

Translation

Translation is the process by which the RNA codes for specific proteins. It is an active process which requires energy. This energy is provided by the charged tRNA molecules. Ribosomes initiate the translation process. The ribosomes consist of a larger subunit and a smaller subunit. The larger subunit, in turn, consists of two tRNA molecules placed close enough so that peptide bond can be formed at the expense of enough energy. The mRNA enters the smaller subunit which is then held by the tRNA molecules of the complementary codon present in the larger subunit. Thus, two codons are held by two tRNA molecules placed close to each other and a peptide bond is formed between them. As this process repeats, long polypeptide chains of amino acids are synthesized.

Genetic Code

Genetic code contains the information of the protein manufactured from RNA. There are basically three nucleotides and four nitrogenous bases, which collectively form a triplet codon that codes for one amino acid. Therefore, the number of possible amino acids ranges to 4 x 4 x 4 = 64 amino acids. There are 20 naturally existing amino acids.

The genetic code degenerates. This was explained by the features of the genetic code, according to which a few amino acids are coded by more than one codon thus causing them to degenerate. Each codon codes for only one specific amino acid and the codes are universal irrespective of the type of organism. Out of the 64 codons, 3 are stop codons which stop the process of transcription and one of the codons is an initiator codon i.e. AUG coding for Methionine.

Gene Expression

Gene expression is the process in which DNA directs the synthesis of functional products, such as proteins. Cells can regulate gene expression at various stages. It allows organisms to generate different cell types and enables cells to adapt to internal and external factors.

Genetic Information Flows from DNA to RNA to Protein

A gene is a stretch of DNA that serves as the blueprint for functional RNAs and proteins. Since DNA is made up of nucleotides and proteins consist of amino acids, a mediator is required to convert the information that is encoded in DNA into proteins. This mediator is the messenger RNA (mRNA). mRNA copies the blueprint from DNA by a process called transcription. In eukaryotes, transcription takes place in the nucleus by complementary base-pairing with the DNA template. The mRNA is then processed and transported into the cytoplasm where it serves as a template for protein synthesis during translation. In prokaryotes, which lack a nucleus, the processes of transcription and translation occur at the same location and almost simultaneously since the newly-formed mRNA is susceptible to rapid degradation.

Gene Expression can be Regulated at any Stage during Transcription

Every cell of an organism contains the same DNA, and consequently the same set of genes. However, not all genes in a cell are "turned on" or use to synthesize proteins. A gene is said to be "expressed" when the protein it encodes is produced by the cell. Gene expression is regulated to ensure proper generation of proteins in specific cells at specific times. Various intrinsic and extrinsic mechanisms regulate gene expression before and during transcription.

The structure of chromatin—compacted DNA and its associated histone proteins—can be chemically modified to be open or closed. Such modifications allow or restrict access of the transcriptional machinery to the DNA. Chromatin modification is an intrinsic mechanism employed during development to form different cell types (e.g., neuron versus muscle cell) from the same genome.

DNA-binding proteins, called transcription factors, regulate transcription by binding to specific DNA sequences near or within the coding regions of genes. Transcription factors that promote the initiation of transcription are called activators. Proteins that prevent the transcription machinery from binding to the transcription initiation site are called repressors. Transcriptional activators or repressors respond to external stimuli such as signaling molecules, nutritional deficiencies, temperature, and oxygen.

Gene Expression can be Post-transcriptionally and Post-translationally Regulated

Gene expression can be regulated by post-transcriptional mRNA processing. In eukaryotes, transcribed mRNA undergoes splicing and other modifications that protect the ends of the RNA strand from degradation. Splicing removes introns—segments that do not encode proteins—and joins together the protein-coding regions called exons. Alternative splicing allows for the expression of functionally diverse proteins from the same gene. Regulation of gene expression by alternative splicing plays an important role in organ development, cell survival and proliferation, and adaptation to environmental factors.

Gene expression can also be altered by regulating the translation of mRNA into proteins. Translation can be regulated by microRNAs—small, non-coding RNAs—that bind to a specific mRNA sequence and block initiation of translation or degrade the transcribed mRNA. In addition, proteins called translational repressors can bind to RNA and interfere with the initiation of translation.

Translated polypeptides undergo processing to form functional proteins. The addition or removal of chemical groups can alter the activity, stability, and localization of proteins in a cell. For instance, the addition or removal of phosphoryl groups ($-PO_3^{2-}$) can activate or inactivate proteins. Similarly, the addition of ubiquitin groups causes protein degradation. Thus, post-translational protein modifications are the final stage of gene regulation.

Gene Expression Regulation

Some genes are expressed continuously, as they produce proteins involved in basic metabolic functions; some genes are expressed as part of the process of cell differentiation; and some genes are expressed as a result of cell differentiation. Mechanisms of gene regulation include:

- Regulating the rate of transcription. This is the most economical method of regulation.
- Regulating the processing of RNA molecules, including alternative splicing to produce more than one protein product from a single gene.
- Regulating the stability of mRNA molecules.
- Regulating the rate of translation.
- Regulation of post translation modification.

Modification of DNA

In eukaryotes, the accessibility of large regions of DNA can depend on its chromatin structure, which can be altered as a result of histone modifications directed by DNA methylation, ncRNA, or DNA-binding protein. Hence these modifications may up or down regulate the expression of a gene. Some of these modifications that regulate gene expression are inheritable and are referred to as epigenetic regulation.

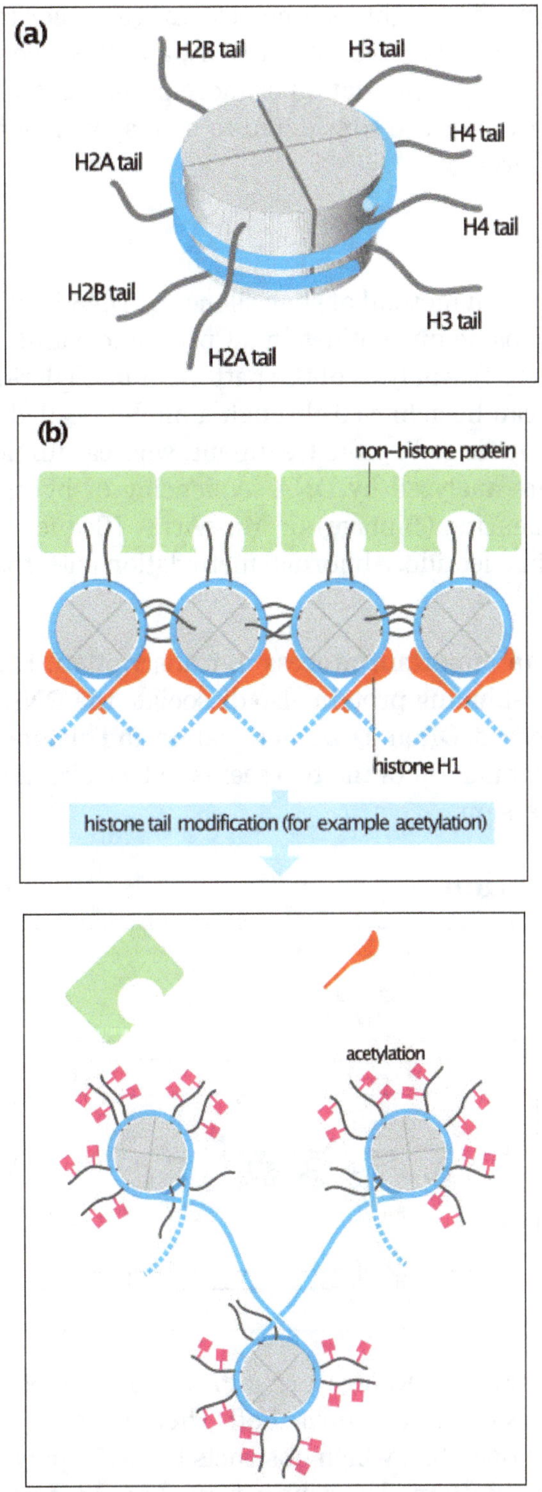

Figure: Histone tails and their function in chromatin formation.

Structural

Transcription of DNA is dictated by its structure. In general, the density of its packing is indicative of the frequency of transcription. Octameric protein complexes called histones together with a segment of DNA wound around the eight histone proteins (together referred to as a nucleosome) are responsible for the amount of supercoiling of DNA, and these complexes can be temporarily modified by processes such as phosphorylation or more permanently modified by processes such as methylation. Such modifications are considered to be responsible for more or less permanent changes in gene expression levels.

Chemical

Methylation of DNA is a common method of gene silencing. DNA is typically methylated by methyltransferase enzymes on cytosine nucleotides in a CpG dinucleotide sequence (also called "CpG islands" when densely clustered). Analysis of the pattern of methylation in a given region of DNA (which can be a promoter) can be achieved through a method called bisulfite mapping. Methylated cytosine residues are unchanged by the treatment, whereas unmethylated ones are changed to uracil. The differences are analyzed by DNA sequencing or by methods developed to quantify SNPs, such as Pyrosequencing (Biotage) or MassArray (Sequenom), measuring the relative amounts of C/T at the CG dinucleotide. Abnormal methylation patterns are thought to be involved in oncogenesis.

Histone acetylation is also an important process in transcription. Histone acetyltransferase enzymes (HATs) such as CREB-binding protein also dissociate the DNA from the histone complex, allowing transcription to proceed. Often, DNA methylation and histone deacetylation work together in gene silencing. The combination of the two seems to be a signal for DNA to be packed more densely, lowering gene expression.

Regulation of Transcription

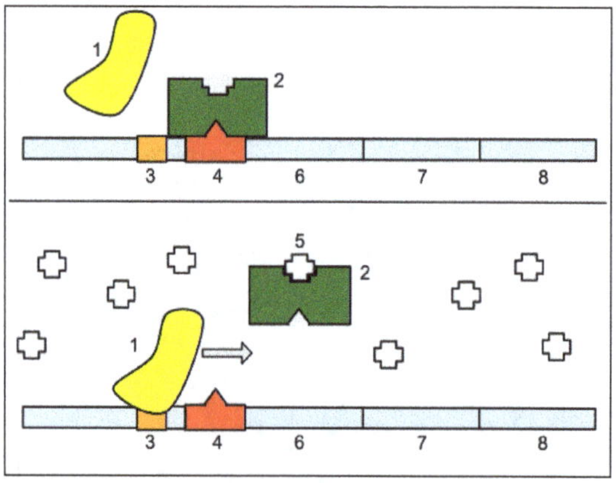

Figure shows 1: RNA Polymerase, 2: Repressor, 3: Promoter, 4: Operator, 5: Lactose, 6: lacZ, 7: lacY, 8: lacA. Top: The gene is essentially turned off. There is no lactose to inhibit the repressor, so the repressor binds to the operator, which obstructs the RNA polymerase from binding to the promoter and making lactase. Bottom: The gene is turned on. Lactose is inhibiting the repressor, allowing the RNA polymerase to bind with the promoter, and express the genes, which synthesize lactase. Eventually, the lactase will digest all of the lactose, until there is none to bind to the repressor. The repressor will then bind to the operator, stopping the manufacture of lactase.

Regulation of transcription thus controls when transcription occurs and how much RNA is created.

Transcription of a gene by RNA polymerase can be regulated by several mechanisms. Specificity factors alter the specificity of RNA polymerase for a given promoter or set of promoters, making it more or less likely to bind to them (i.e., sigma factors used in prokaryotic transcription). Repressors bind to the Operator, coding sequences on the DNA strand that are close to or overlapping the promoter region, impeding RNA polymerase's progress along the strand, thus impeding the expression of the gene. The image to the right demonstrates regulation by a repressor in the lac operon. General transcription factors position RNA polymerase at the start of a protein-coding sequence and then release the polymerase to transcribe the mRNA. Activators enhance the interaction between RNA polymerase and a particular promoter, encouraging the expression of the gene. Activators do this by increasing the attraction of RNA polymerase for the promoter, through interactions with subunits of the RNA polymerase or indirectly by changing the structure of the DNA. Enhancers are sites on the DNA helix that are bound by activators in order to loop the DNA bringing a specific promoter to the initiation complex. Enhancers are much more common in eukaryotes than prokaryotes, where only a few examples exist (to date). Silencers are regions of DNA sequences that, when bound by particular transcription factors, can silence expression of the gene.

Regulation of Transcription in Cancer

In vertebrates, the majority of gene promoters contain a CpG island with numerous CpG sites. When many of a gene's promoter CpG sites are methylated the gene becomes silenced. Colorectal cancers typically have 3 to 6 driver mutations and 33 to 66 hitchhiker or passenger mutations. However, transcriptional silencing may be of more importance than mutation in causing progression to cancer. For example, in colorectal cancers about 600 to 800 genes are transcriptionally silenced by CpG island methylation. Transcriptional repression in cancer can also occur by other epigenetic mechanisms, such as altered expression of microRNAs. In breast cancer, transcriptional repression of BRCA1 may occur more frequently by over-expressed microRNA-182 than by hypermethylation of the BRCA1 promoter.

Regulation of Transcription in Addiction

One of the cardinal features of addiction is its persistence. The persistent behavioral changes appear to be due to long-lasting changes, resulting from epigenetic alterations affecting gene expression, within particular regions of the brain. Drugs of abuse cause three types of epigenetic alteration in the brain. These are (1) Histone acetylations and histone methylations, (2) DNA methylation at CpG sites, and (3) Epigenetic downregulation or upregulation of microRNAs.

Chronic nicotine intake in mice alters brain cell epigenetic control of gene expression through acetylation of histones. This increases expression in the brain of the protein FosB, important in addiction. Cigarette addiction was also studied in about 16,000 humans, including never smokers, current smokers, and those who had quit smoking for up to 30 years. In blood cells, more than 18,000 CpG sites (of the roughly 450,000 analyzed CpG sites in the genome) had frequently altered methylation among current smokers. These CpG sites occurred in over 7,000 genes, or roughly a third of known human genes. The majority of the differentially methylated CpG sites returned to the level of never-smokers within five years of smoking cessation. However, 2,568 CpGs among 942 genes remained differentially methylated in former versus never smokers. Such remaining epigenetic changes can be viewed as "molecular scars" that may affect gene expression.

In rodent models, drugs of abuse, including cocaine, methampheamine, alcohol and tobacco smoke products, all cause DNA damage in the brain. During repair of DNA damages some individual repair events can alter the methylation of DNA and/or the acetylations or methylations of histones at the sites of damage, and thus can contribute to leaving an epigenetic scar on chromatin. Such epigenetic scars likely contribute to the persistent epigenetic changes found in addiction.

Regulation of Transcription in Learning and Memory

In mammals, methylation of cytosine in DNA is a major regulatory mediator. Methylated cytosines primarily occur in dinucleotide sequences where cytosine is followed by a guanine, a CpG site. The total number of CpG sites in the human genome is approximately 28 million and generally about 70% of all CpG sites have a methylated cytosine.

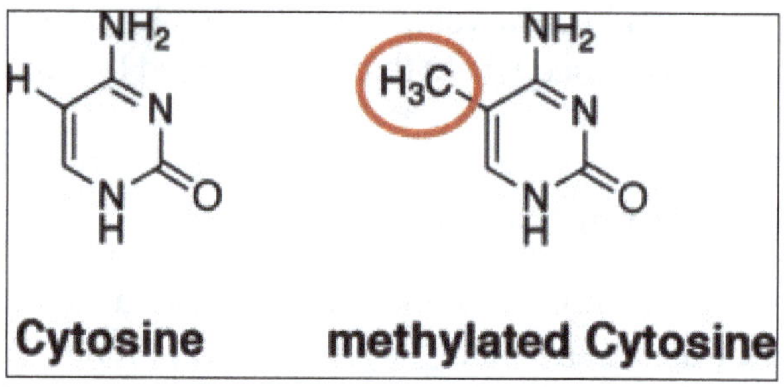

Figure: DNA methylation is the addition of a methyl group to the DNA that happens at cytosine. The image shows a cytosine single ring base and a methyl group added on to the 5 carbon. In mammals, DNA methylation occurs almost exclusively at a cytosine that is followed by a guanine.

In a rat, a painful learning experience, contextual fear conditioning, can result in a life-long fearful memory after a single training event. Cytosine methylation is altered in the promoter regions of about 9.17% of all genes in the hippocampus neuron DNA of a rat that has been subjected to a brief fear conditioning experience. The hippocampus is where new memories are initially stored.

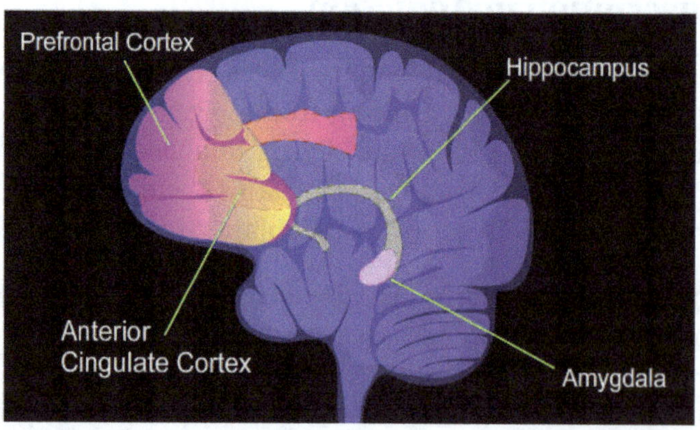

Figure: The identified areas of the human brain are involved in memory formation.

Methylation of CpGs in a promoter region of a gene represses transcription while methylation of CpGs in the body of a gene increases expression. TET enzymes play a central role in demethylation of methylated cytosines. Demethylation of CpGs in a gene promoter by TET enzyme activity increases transcription of the gene.

When contextual fear conditioning is applied to a rat, more than 5,000 differentially methylated regions (DMRs) (of 500 nucleotides each) occur in the rat hippocampus neural genome both one hour and 24 hours after the conditioning in the hippocampus. This causes about 500 genes to be up-regulated (often due to demethylation of CpG sites in a promoter region) and about 1,000 genes to be down-regulated (often due to newly formed 5-methylcytosine at CpG sites in a promoter region). The pattern of induced and repressed genes within neurons appears to provide a molecular basis for forming the first transient memory of this training event in the hippocampus of the rat brain.

Post-transcriptional Regulation

After the DNA is transcribed and mRNA is formed, there must be some sort of regulation on how much the mRNA is translated into proteins. Cells do this by modulating the capping, splicing, addition of a Poly(A) Tail, the sequence-specific nuclear export rates, and, in several contexts, sequestration of the RNA transcript. These processes occur in eukaryotes but not in prokaryotes. This modulation is a result of a protein or transcript that, in turn, is regulated and may have an affinity for certain sequences.

Three Prime Untranslated Regions and MicroRNAs

Three prime untranslated regions (3'-UTRs) of messenger RNAs (mRNAs) often contain regulatory sequences that post-transcriptionally influence gene expression. Such 3'-UTRs often contain both binding sites for microRNAs (miRNAs) as well as for regulatory proteins. By binding to specific sites within the 3'-UTR, miRNAs can decrease gene expression of various mRNAs by either inhibiting translation or directly causing degradation of the transcript. The 3'-UTR also may have silencer regions that bind repressor proteins that inhibit the expression of a mRNA.

The 3'-UTR often contains miRNA response elements (MREs). MREs are sequences to which miRNAs bind. These are prevalent motifs within 3'-UTRs. Among all regulatory motifs within the 3'-UTRs (e.g. including silencer regions), MREs make up about half of the motifs.

As of 2014, the miRBase web site, an archive of miRNA sequences and annotations, listed 28,645 entries in 233 biologic species. Of these, 1,881 miRNAs were in annotated human miRNA loci. miRNAs were predicted to have an average of about four hundred target mRNAs (affecting expression of several hundred genes). Freidman et al. estimate that >45,000 miRNA target sites within human mRNA 3'-UTRs are conserved above background levels, and >60% of human protein-coding genes have been under selective pressure to maintain pairing to miRNAs.

Direct experiments show that a single miRNA can reduce the stability of hundreds of unique mRNAs. Other experiments show that a single miRNA may repress the production of hundreds of proteins, but that this repression often is relatively mild (less than 2-fold). The effects of miRNA dysregulation of gene expression seem to be important in cancer. For instance, in gastrointestinal cancers, a 2015 paper identified nine miRNAs as epigenetically altered and effective in down-regulating DNA repair enzymes.

The effects of miRNA dysregulation of gene expression also seem to be important in neuropsychiatric disorders, such as schizophrenia, bipolar disorder, major depressive disorder, Parkinson's disease, Alzheimer's disease and autism spectrum disorders.

Regulation of Translation

The translation of mRNA can also be controlled by a number of mechanisms, mostly at the level of initiation. Recruitment of the small ribosomal subunit can indeed be modulated by mRNA secondary structure, antisense RNA binding, or protein binding. In both prokaryotes and eukaryotes, a large number of RNA binding proteins exist, which often are directed to their target sequence by the secondary structure of the transcript, which may change depending on certain conditions, such as temperature or presence of a ligand (aptamer). Some transcripts act as ribozymes and self-regulate their expression.

Examples of Gene Regulation

- Enzyme induction is a process in which a molecule (e.g., a drug) induces (i.e., initiates or enhances) the expression of an enzyme.
- The induction of heat shock proteins in the fruit fly *Drosophila melanogaster*.
- The Lac operon is an interesting example of how gene expression can be regulated.
- Viruses, despite having only a few genes, possess mechanisms to regulate their gene expression, typically into an early and late phase, using collinear systems regulated by anti-terminators (lambda phage) or splicing modulators (HIV).
- Gal4 is a transcriptional activator that controls the expression of GAL1, GAL7, and GAL10 (all of which code for the metabolic of galactose in yeast). The GAL4/UAS system has been used in a variety of organisms across various phyla to study gene expression.

Developmental Biology

A large number of studied regulatory systems come from developmental biology. Examples include:

- The colinearity of the Hox gene cluster with their nested antero-posterior patterning.
- Pattern generation of the hand (digits - interdigits): the gradient of sonic hedgehog (secreted inducing factor) from the zone of polarizing activity in the limb, which creates a gradient of active Gli3, which activates Gremlin, which inhibits BMPs also secreted in the limb, results in the formation of an alternating pattern of activity as a result of this reaction-diffusion system.
- Somitogenesis is the creation of segments (somites) from a uniform tissue (Pre-somitic Mesoderm). They are formed sequentially from anterior to posterior. This is achieved in amniotes possibly by means of two opposing gradients, Retinoic acid in the anterior (wavefront) and Wnt and Fgf in the posterior, coupled to an oscillating pattern (segmentation clock) composed of FGF + Notch and Wnt in antiphase.
- Sex determination in the soma of a Drosophila requires the sensing of the ratio of autosomal genes to sex chromosome-encoded genes, which results in the production of sexless splicing factor in females, resulting in the female isoform of doublesex.

Circuitry

Up-regulation and Down-regulation

Up-regulation is a process that occurs within a cell triggered by a signal (originating internal or external to the cell), which results in increased expression of one or more genes and as a result the proteins encoded by those genes. Conversely, down-regulation is a process resulting in decreased gene and corresponding protein expression.

- Up-regulation occurs, for example, when a cell is deficient in some kind of receptor. In this case, more receptor protein is synthesized and transported to the membrane of the cell and, thus, the sensitivity of the cell is brought back to normal, re-establishing homeostasis.
- Down-regulation occurs, for example, when a cell is overstimulated by a neurotransmitter, hormone, or drug for a prolonged period of time and the expression of the receptor protein is decreased in order to protect the cell.

Inducible Vs. Repressible Systems

Gene Regulation can be summarized by the response of the respective system:

- Inducible systems - An inducible system is off unless there is the presence of some molecule (called an inducer) that allows for gene expression. The molecule is said to "induce expression". The manner by which this happens is dependent on the control mechanisms as well as differences between prokaryotic and eukaryotic cells.
- Repressible systems - A repressible system is on except in the presence of some molecule (called a corepressor) that suppresses gene expression. The molecule is said to "repress expression". The manner by which this happens is dependent on the control mechanisms as well as differences between prokaryotic and eukaryotic cells.

The GAL4/UAS system is an example of both an inducible and repressible system. Gal4 binds an upstream activation sequence (UAS) to activate the transcription of the GAL1/GAL7/GAL10 cassette. On the other hand, a MIG1 response to the presence of glucose can inhibit GAL4 and therefore stop the expression of the GAL1/GAL7/GAL10 cassette.

Theoretical Circuits

- Repressor/Inducer: An activation of a sensor results in the change of expression of a gene
- Negative feedback: The gene product downregulates its own production directly or indirectly, which can result in:
  - Keeping transcript levels constant/proportional to a factor.
  - Inhibition of run-away reactions when coupled with a positive feedback loop.
  - Creating an oscillator by taking advantage in the time delay of transcription and translation, given that the mRNA and protein half-life is shorter.

- Positive feedback: The gene product upregulates its own production directly or indirectly, which can result in:
    - Signal amplification.
    - Bistable switches when two genes inhibit each other and both have positive feedback.
    - Pattern generation.

Gene Expression in Eukaryotes

The simplest meaning of the gene expression is the phenotypic expression of the genes via transcription and translation processes. These are the events by which the functional product usually protein is formed by the transfer the information content of the gene. There are some genes like rRNAs, tRNAs and small RNAs encoding genes which resulted in RNA as a functional product. Gene expression is more complex in case of eukaryotes as compared to the prokaryotes. Eukaryotic cells have the DNA within the nucleus of the cell where its transcription takes place and transcribed into mRNA. Then the mRNA is transported to the cytoplasm from the nucleus and ultimately the proteins are synthesized through the translation of mRNA by the ribosomes. It is the nuclear membrane that physically separates both the process i.e. transcription and translation. It is the nucleus where transcription occurs and translation occurs in the cytoplasm. The gene regulation occurs at all the levels of the process. These are epigenetic level, transcriptional level, post-transcriptional level, translational level and post-translational level. The epigenetic level is unwinding of the DNA and loosened from nucleosomes and leads to the binding to the transcription factors. In the transcriptional level the RNA is transcribed from the DNA. Both the levels occur in the nucleus of the cell. The post transcriptional level is processing of the RNA and its transportation to the cytoplasm after transcription. Whereas the translation level is the translation of the RNA to protein and finally the post-translational level when finally the functional protein is formed.

There are some advantages in the cloned gene expression in the eukaryotes. These organisms have the ability to produce the post-translational modifications in proteins. Their expression system produces stable and functional proteins which are biologically very active. As compared to the eukaryotes, for the cloned gene expression in prokaryotes, bacteria are an ideal host for DNA amplification which can produce various prokaryotic and eukaryotic proteins. But there are some drawbacks of this system of expression of cloned genes i.e. to produce post-translational modifications bacteria do not produce essential enzymes for such modifications like polypeptide cleavage and binding of the carbohydrate units. So, only eukaryotic cells can be the better host for the expression of the eukaryotic genes.

Techniques for the Manipulation of Gene Expression in Eukaryotes

New techniques have been developed in the field of genetic engineering for the manipulation of the gene expression by the scientists. RNA interference is one of them and is most highlighted. Besides this technology two other technologies have also been progressed. These are antisense oligonucleotide technology and designed transcription factors. The resurrection of the antisense oligonucleotide technology takes placed after the chemical modification of these molecules. On the other hand the designed transcription factors evolved as the powerful and most suitable technique for the study of the up-regulation and down-regulation of the genes.

RNA Interference (RNAi)

This technology is the most highlighted and acceptable technique to study the endogenous gene functions in different organisms and to explore their therapeutic strategies. With the help of this technology phenotypic knockdowns of most of the genes in the mammalian cells can be generated. This technology was first developed to study the post-translational induction mechanism and RNA degradation of the specific sequences by the long dsRNAs introduction into worms. This type of silencing machineries is extremely conserved in various organisms as well as in mammals that uses dsRNAs to knockdown the expression of the target genes. Biochemically, during the process of RNA silencing, smaller dsRNAs of the sizes of 20-30 nucleotides with 2nt 3′ overhangs has been cleaved from the long double stranded RNAs. Which are called as small interfering RNAs (siRNAs) whose one strand is then integrated to form a complex with a protein named Argonaute protein? This protein acts as the core protein component of the complex. Which leads to the inactivation of the gene?

Mechanism of RNAi Silencing

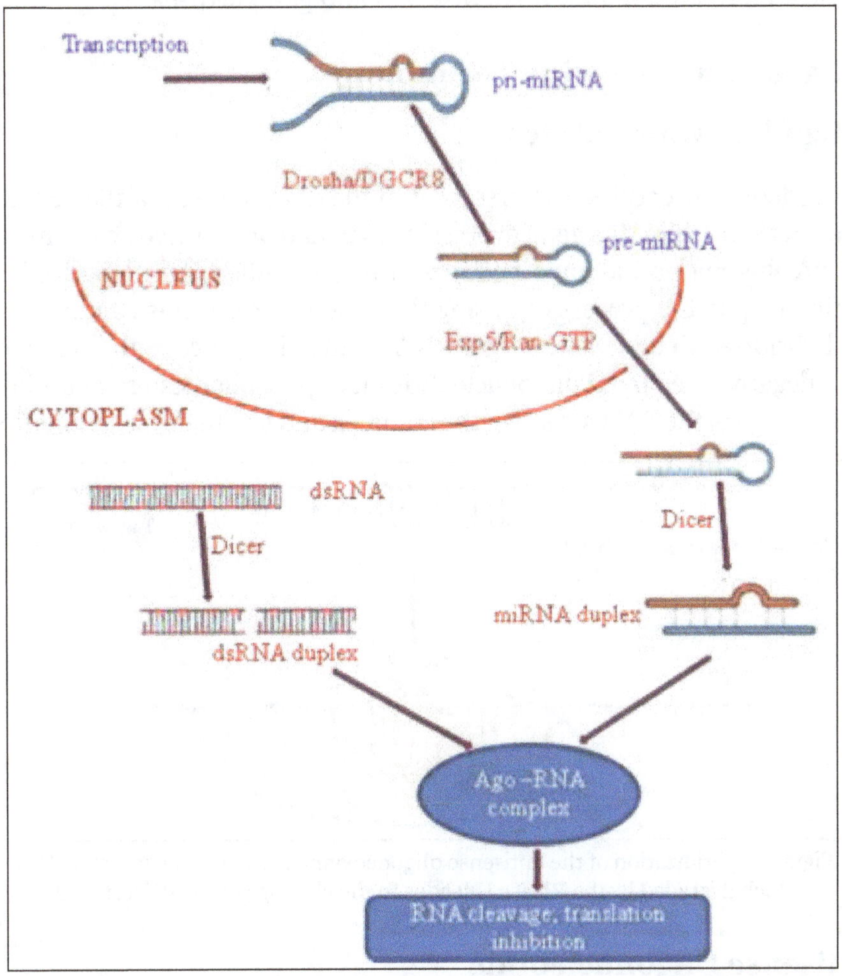

Figure: miRNAs and siRNAs production and their function in animal cells.

In the conventional RNAi silencing mechanism, the small RNA hybridizes into RNA substrates and guides the Ago protein in mammals leads to the cleavage of the target RNA in the core of the hybrid. This complex (endoribonuclease) is known as RNA-induced silencing complex (RISC). The other nuclease further degrades the cleaved RNA in the cell. In the other pathway, which is termed as inhibition of translation? In this process, recognition of the 3' UTR regions of the targeted mRNA by the small RNA takes place but due to the low sequence complementarities between the RNAs, Ago-mediated, rarely these specific mRNA cleavages occurs. Alternatively, the Ago complex and the mRNA interaction inhibits the mRNA translation, even though the whole mechanism is not very clear.

In the latest method of RNAi silencing mechanism, the transcription is suppressed by the Ago protein and RNA complex by mediating the chromatin structure modifications. The former two methods demonstrate the RNA silencing in mammals. While long dsRNAs bring out strong RNAi effects in other animals like invertebrates, whereas in case of most of the vertebrates the dsRNAs do not show any effect. The initiation of RNAi by the siRNAs is proved to be a a major breakthrough and provides a better method to study of gene expression in mammalian systems. In contrast to the techniques like gene targeting, where we can produce knockout mice and the approach is very time consuming and very costly, the RNAi technology is more faster and can be applied to both the mammalian cell cultures as well as in animals. This technology has also been developed for therapeutic purposes for the treatment of human diseases and genetic diseases.

Antisense Oligodeoxynucleotide Technology

Antisense Oligodeoxynucleotides

The antisense oligodeoxynucleotides was first used in the development of the actual antisense technology in the year 1978 to inhibit the gene expression. Till date in the studies of antisense a variety of both DNA and RNA oligonucleotides modified as well as unmodified of the length of 18nt to 25nt have been used. The first step in the process of raising the gene knockdown is common in all these oligonucleotides and hybridized to their cell's target RNAs. After this process, the cellular enzyme RNase H recognizes the negatively charged oligonucleotides like phosphodiesters and phosphorothioates. These results in cleavage of the RNA strand of the complex and ultimately lead to mRNA degradation.

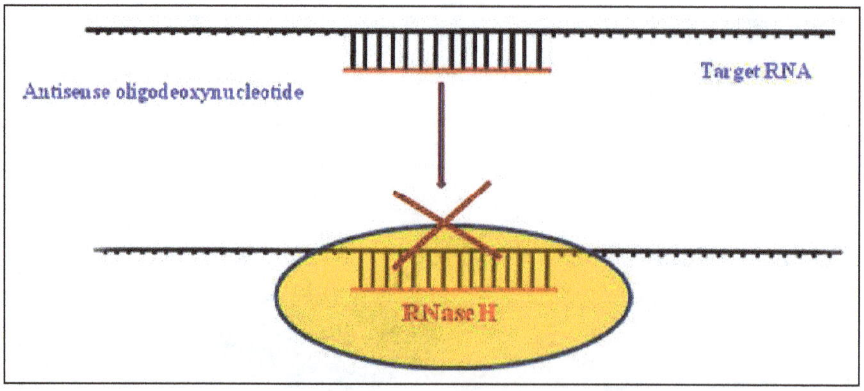

Figure: Hybridization of the antisense oligodeoxynucleotide to the target mRNA which degraded by the RNase H leades to the cleavage of the RNA strand.

Modified Antisense Oligonucleotide

The other class of the antisense molecules which inactivates the cellular enzyme RNase H. Due to the nature of the formation of the duplex with the target RNA, it is the steric hindrance that causes translation inhibition or it interferes with splicing of pre-mRNA. This class of the antisense molecules are modified derivatives of the nucleic acids such as morpholinos, 2'-O-methyls, 2'-O-allyls, locked nucleic acids and peptide nucleic acids.

Figure: Hybridization of the target mRNA with the chemically modified antisense oligonucleotide inhibits translation by steric hindrance.

Ribozymes or Deoxyribozymes

The other categories of sequence specific gene silencing molecules are known as ribozymes, which catalyzes the chemical reactions in the absence of the protein co-factors. For the application of the gene-knockdown, ribozymes that are flanked by the sequences complementary to the target RNA are used. For the gene silencing, the ribozymes binds to the RNA via Watson-Crick base pairing that leads to the breakdown of the phosphodiester linkages of the RNA target. As soon as the target RNA cleaves, dissociation of the ribozymes occurs and consequently repeatition of the cleavage can occurs on additional substrates. The ribozyme mostly used in the field of molecular biology and biotechnology is hammerhead ribozyme, isolated from viroid RNAs. During replication it undergoes sitespecific self-cleavage. Hammerhead ribozyme can be transformed from a cis-cleaving RNA enzyme to a target-specific trans-cleaving molecule by separating the catalytic and substrate strands of the ribozyme. It has an advantage of nuclease resistance, increased activity in physiological concentrations of Mg2+, and improvement of accessibility to the target sequences.

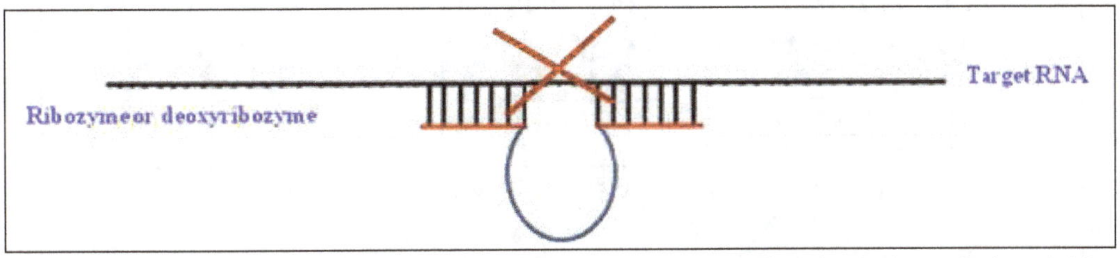

Figure: Cleavage of the target RNA by the ribozymes directly due to their intrinsic catalytic activities.

The catalytic DNAs or deoxyribozymes are the other types of the nucleic acid-based agents that can be used for the gene inhibition. It binds to their substrates of RNA via Watson Crick base pairing and cleaves the target RNA, like that of ribozymes works. These are produced by in vitro and have catalytic motifs of 8–17 and 10–23 with different cleavage site specificities.

Designed Transcription Factors

The transcription factors are the protein that contains DNA-binding domains that specifically recognizes the base sequences and helps in transcriptional activation and repression. The transcription factors and chromatin forms a protein complex that acts as coactivators or corepressors. The coactivators such as CBP–p300 complex activates transcription along with the modification of histone, the SWI–SNF chromatin remodeling complex, and the mediator complex links transcription factors to the basal transcription process. However, the corepressors like Sin3 and NuRD complexes that constitutes histone deacetylases converts nucleosomes to a incompetent state of transcription. With the help of the screening library process new transcription factors can be created with some novel DNA- binding domains. Which can be used to generate functional designed TFs by their combination with the known transactivating or repressor domains. The domains from the herpes virus VP16 protein or from the NFκB p65 subunit are the commonly used transactivators and the domain from Krupple-associated box and mSin3 interaction are commonly used as repressors. The Cys2–His2 zinc finger (Zif) has been used as a DNA binding entity to designed TFs. The characteristics of the C2H2 Zif includes the stabilization by the zinc ion, constituted of about 30 amino acids, and these are arranged as two β sheets and an α-helix, with the helix being the primary DNA recognition moiety. As the relationship between protein structures and base recognition has already been studied extensively and based on the modular nature of Zif–DNA interaction the C2H2 Zif is utilized to design the transcription factors. So, with the start of the the α-helix the amino acids at positions −1 to 6 fits into the major groove of the DNA and mediated by 3' base motif recognition. The residue at positions −1, 3, and 6 interacts directly with the DNA strand triplet bases at the position of 3', central, and 5'. While the interaction between the additional base and backbone can occurred. The polydactyl multi-Zif proteins is allowed to create during this process that can recognizes even long DNA stretches which is emerged as a new tool for gene manipulation.

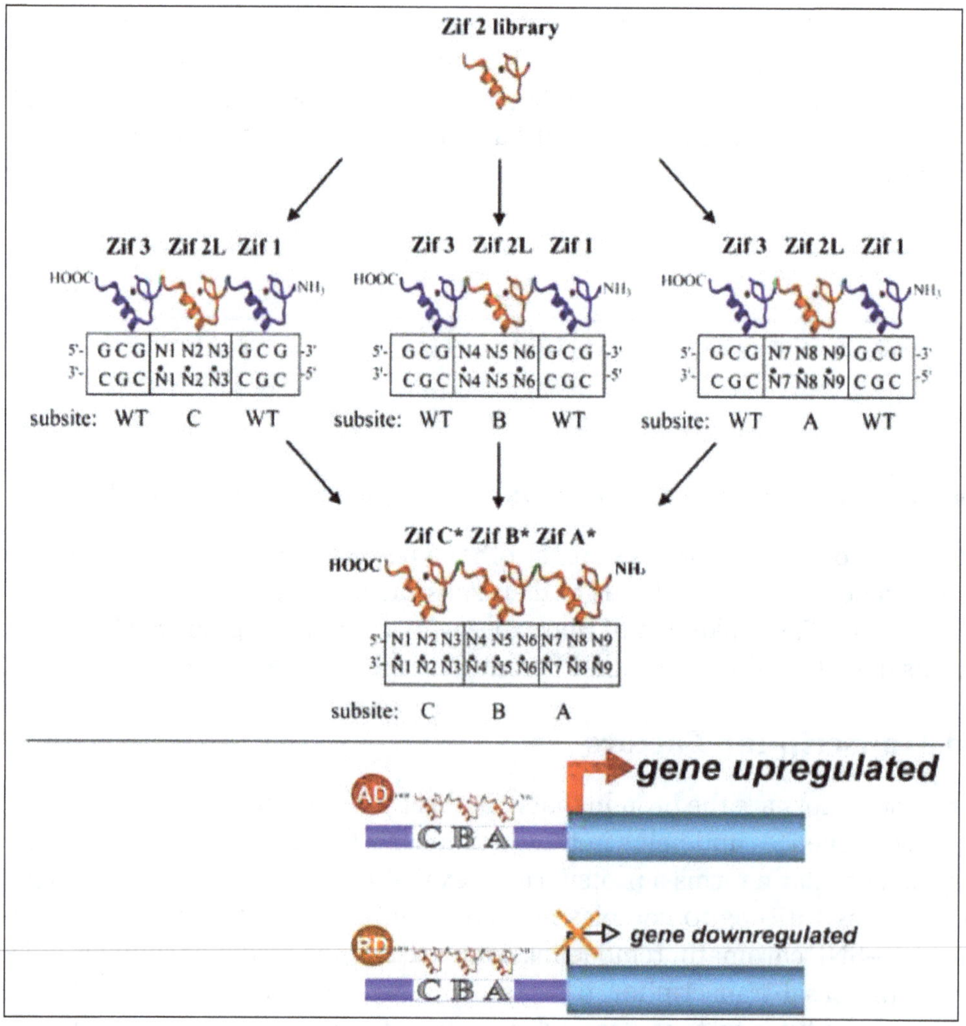

During the last few years designed TFs has been extensively studied and the large repertories of C2H2 Zifs with known binding specificities was created. Libraries of libraries have been created for the modified Zifs targeting multiple three base sites which further can be assembled into multi-Zif TFs. These libraries can then be transfected into mammalian cells for the screening of the phenotypic expression of a particular gene and also comparable to the the siRNA libraries. An important advantage of this technology is that once the identification of the desirable Zif-TF is done, its expression in cells can be regulated in a variety of ways.

Figure shows stretagy to develop designed transcription factors (TFs). First step is to select Zincfinger domains (Zifs) which binds desired DNA triplet. Multi-Zifs complex is formed by the combination of the Individual Zifs followed by the selective binding to the extended regions of DNA in the promoter regions of genes. The upregulation and down regulation of the gene expression is occurred with the addition of transactivating or repressor domains. In the above figue N1–N9 are different bases and N* is the complement. The A–C denotes triplets. Zif* is the Zinc-finger domains selected from screening of the library. WT indicate wild type. AD and RD indicate activation domain and repressor domain respectively.

Gene Expression in Prokaryotes

Prokaryotes only transcribe genes that their end-proteins are needed at the time. They do this in order to save up energy and increase efficiency. The regulation of gene expression is depended mainly on their immediate environment, for example on the presence and absence of nutrients. Gene expression in prokaryotes occurs primarily at the level of trascription.

What are Operons?

Bacteria genes that encode for proteins with closely related functions are found grouped along with cis-acting regulatory elements that determine the transcription of these genes, thus these genes are regulated in a coordinated way. These clusters of genes are called operons, and their ranscription product is a single polycistronic mRNA. Organization of genes in operons contributes to the regulation of gene expression. The operon can therefore be categorized as inducible or repressible.

Inducible operons:

- They include genes that encode for enzymes that take part in metabolic pathways an the expression of the gene is controlled by the substrate.
- Example: the "Lac Operon".

Repressible operons:

- They include genes that encode for enzymes involved in biosynthetic pathways, and the expression of the gene is controlled by the end-product of the pathway.
- Example: the "Trp Operon".

What are Operators?

These are segments of DNA that regulate the activity of the structural genes of the operon:

- If the operator is not bound by a repressor molecule, then the RNA polymerase can pass over the operator and move to the protein coding genes.
- If the operator, on the other hand, is bound by a repressor molecule, then the RNA polymerase is blocked behind the repressor molecule.
- However, when an "Inducer molecule" is present, it binds the repressor molecule causing it to change shape, rendering it incapable of binding the operator. The RNA polymerase moves freely.

The Lactose Operon

- The lactose operon contains 3 genes, Lac Z, Y and A. These genes encode for enzymes required to metabolize lactose → beta-galactosidase, lactose permease, and beta-galactoside transacetylase.
- Another regulatory gene, Lac I, is expressed separately and lies upstream of the operon. This gene ecodes for the lac repressor which regulates the expression of Lac Z, Y ans A.
- In the absence of lactose:
    - The lac repressor bind to a DNA sequence called the "operator" (found between the lac Z gene and the lac promoter).
    - In this way the lac repressor blocks the path of RNA polymerase to reach the lac Z,Y and A genes → operon remains switched off.
- In the presence of lactose:
    - Lactose molecules are metabolized by the lac enzymes, an intermediate is formed called allolactose (an isomer of lactose).
    - Allolactose acts as an inducer by binding to the lactose repressor and changing its conformation → therefore it can no longer bind to the operator.

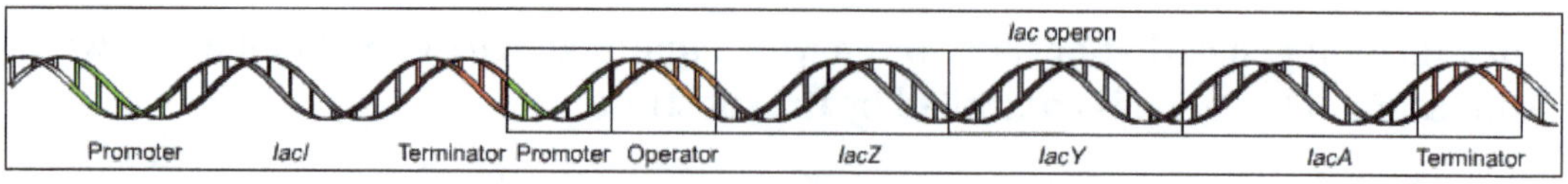

- In the presence of glucose (even if present together with lactose):
    - Glucose is preferred because it requires less energy expenditure to be broken down.
    - The operon senses the glucose presence and by a mechanism called "catabolite repression" it is switched off.
    - The regulatory protein "CAP" (Catabolite Activator Protein) binds to a DNA sequence

upstream to the lac promoter and enhances binding of the RNA polymerase leading to an increased transcription of the operon.

- However, CAP will only bind behind the promoter only if cAMP is bound on it. Adenylate cyclase though, the enzyme required for the synthesis of cAMP, is inhibited by glucose and the operon will be eventually expressed at very low rate.
- The oposite also applies.

The Tryptophan operon (Trp operon)

- It includes 5 genes involved in Tryptophan synthesis.
- The genes are expressed as a single mRNA strand, transcribed from an upstream promoter.
- Another regulatory gene encodes for a trp repressor.
- There is also the trp operator that is found just downstream from the promoter.
- If tryptophan is present it will bind the trp repressor. This enables it to bind the operator and block the RNA polymerase. This is a model of end-product inhibition, since tryptophan is the end product of tryptophan biosynthesis.

Attenuation

- Alternatively another method of suppressing gene expression is "attenuation". This is only possible in prokaryotes since t has to do with the fact that the mRNA is translated as soon as it is being transcribed.
- The trp mRNA can form 2 stem-looped structures between the trp-promoter and the 1st trp gene. The large stem loop does not influence transcription and occurs upstream of the smaller stem loop which acts as a transcription terminator. If the large stem loop is not present then the short stem-loop will be formed.
- The formation of the stem-loops on the mRNA depends on the binding of ribosomes to the mRNA.
- Just prior to the stem-loop region of the mRNA, there is an open reading frame of 14 codons, of which 2 of them code for tryptophan.
- If the tryptophan is present in adequate levels in the cytoplasm, it will be translated and the ribosome will keep translocating following closely behind the RNA polymerase preventing the formation of the large loop structure, allowing the terminator structure to form further on and transcription ends.
- If tryptophan is lacking, the ribosome will be stalled as it tries to translate the coding region. This will allow the RNA polymerase to move ahead and leave enough space for the larger stem-loop structure to form, thus allowing the transcription to continue.
- Overall the trp repressor determines whether to switch on or off the entire operon while the method of attenuation determines the efficiency of transcription, according to the level of tryptophan available.

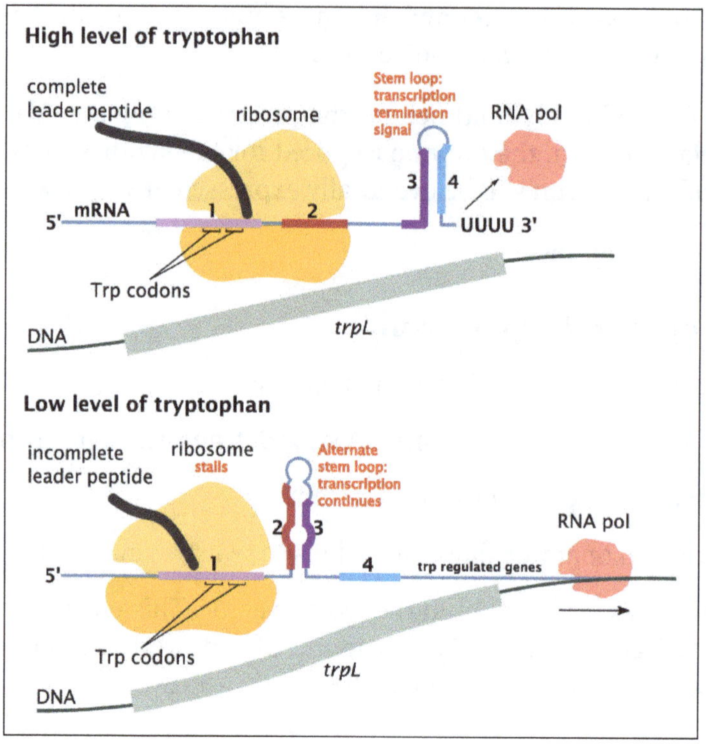

Figure: Trp operon attenuation.

Gene Cloning

Gene cloning is the production of exact copies of a particular gene or DNA sequence using genetic engineering techniques is called gene cloning. The term "gene cloning," "DNA cloning," "molecular cloning," and "recombinant DNA technology" all refer to same technique. When DNA is extracted from an organism, all its genes are obtained. In gene (DNA) cloning a particular gene is copied forming "clones". Cloning is one method used for isolation and amplification of gene of interest. DNA cloning can be achieved by two different methods:

- Cell based DNA cloning.
- Cell-free DNA cloning (PCR).

Requirements for Gene Cloning (Cell-based)

- DNA fragment containing the desired genes to be cloned.
- Restriction enzymes and ligase enzymes.
- Vectors – to carry, maintain and replicate cloned gene in host cell.
- Host cell – in which recombinant DNA can replicate.

Molecular Cloning

Cloning allows for the creation of multiple copies of genes, expression of genes, and study of specific genes. To get the DNA fragment into a bacterial cell in a form that will be copied or expressed, the fragment is first inserted into a plasmid. A plasmid (also called a vector in this context) is a small circular DNA molecule that replicates independently of the chromosomal DNA in bacteria. In cloning, the plasmid molecules can be used to provide a "vehicle" in which to insert a desired DNA fragment. Modified plasmids are usually reintroduced into a bacterial host for replication. As the bacteria divide, they copy their own DNA (including the plasmids). The inserted DNA fragment is copied along with the rest of the bacterial DNA. In a bacterial cell, the fragment of DNA from the human genome (or another organism that is being studied) is referred to as foreign DNA to differentiate it from the DNA of the bacterium (the host DNA).

Plasmids occur naturally in bacterial populations (such as *Escherichia coli*) and have genes that can contribute favorable traits to the organism, such as antibiotic resistance (the ability to be unaffected by antibiotics). Plasmids have been highly engineered as vectors for molecular cloning and for the subsequent large-scale production of important molecules, such as insulin. A valuable characteristic of plasmid vectors is the ease with which a foreign DNA fragment can be introduced. These plasmid vectors contain many short DNA sequences that can be cut with different commonly available restriction enzymes. Restriction enzymes (also called restriction endonucleases) recognize specific DNA sequences and cut them in a predictable manner; they are naturally produced by bacteria as a defense mechanism against foreign DNA. Many restriction enzymes make staggered cuts in the two strands of DNA, such that the cut ends have a 2- to 4-nucleotide single-stranded overhang. The sequence that is recognized by the restriction enzyme is a four- to eight-nucleotide sequence that is a palindrome. Like with a word palindrome, this means the sequence reads the same forward and backward. In most cases, the sequence reads the same forward on one strand and backward on the complementary strand. When a staggered cut is made in a sequence like this, the overhangs are complementary.

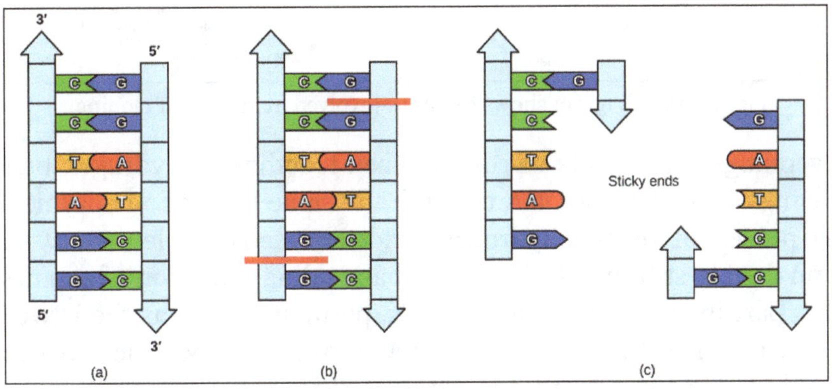

Figure shows in this (a) six-nucleotide restriction enzyme recognition site, notice that the sequence of six nucleotides reads the same in the 5′ to 3′ direction on one strand as it does in the 5′ to 3′ direction on the complementary strand. This is known as a palindrome. (b) The restriction enzyme makes breaks in the DNA strands, and (c) the cut in the DNA results in "sticky ends". Another piece of DNA cut on either end by the same restriction enzyme could attach to these sticky ends and be inserted into the gap made by this cut.

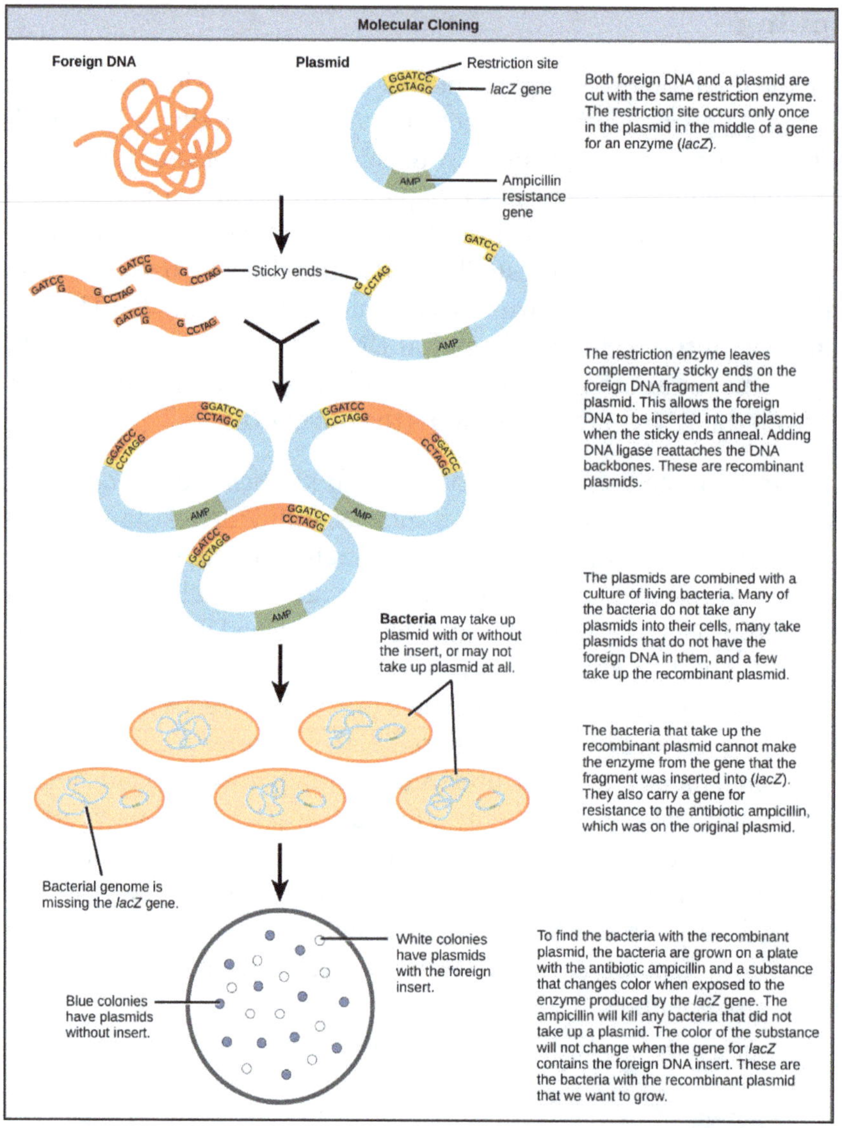

Figure: This diagram shows the steps involved in molecular cloning.

Because these overhangs are capable of coming back together by hydrogen bonding with complementary overhangs on a piece of DNA cut with the same restriction enzyme, these are called "sticky ends." The process of forming hydrogen bonds between complementary sequences on single strands to form double-stranded DNA is called annealing. Addition of an enzyme called DNA ligase, which takes part in DNA replication in cells, permanently joins the DNA fragments when the sticky ends come together. In this way, any DNA fragment can be spliced between the two ends of a plasmid DNA that has been cut with the same restriction enzyme.

Plasmids with foreign DNA inserted into them are called recombinant DNA molecules because they contain new combinations of genetic material. Proteins that are produced from recombinant DNA molecules are called recombinant proteins. Not all recombinant plasmids are capable of expressing genes. Plasmids may also be engineered to express proteins only when stimulated by certain environmental factors, so that scientists can control the expression of the recombinant proteins.

Reproductive Cloning

Reproductive cloning is a method used to make a clone or an identical copy of an entire multicellular organism. Most multicellular organisms undergo reproduction by sexual means, which involves the contribution of DNA from two individuals (parents), making it impossible to generate an identical copy or a clone of either parent. Recent advances in biotechnology have made it possible to reproductively clone mammals in the laboratory.

Natural sexual reproduction involves the union, during fertilization, of a sperm and an egg. Each of these gametes is haploid, meaning they contain one set of chromosomes in their nuclei. The resulting cell, or zygote, is then diploid and contains two sets of chromosomes. This cell divides mitotically to produce a multicellular organism. However, the union of just any two cells cannot produce a viable zygote; there are components in the cytoplasm of the egg cell that are essential for the early development of the embryo during its first few cell divisions. Without these provisions, there would be no subsequent development. Therefore, to produce a new individual, both a diploid genetic complement and an egg cytoplasm are required. The approach to producing an artificially cloned individual is to take the egg cell of one individual and to remove the haploid nucleus. Then a diploid nucleus from a body cell of a second individual, the donor, is put into the egg cell. The egg is then stimulated to divide so that development proceeds. This sounds simple, but in fact it takes many attempts before each of the steps is completed successfully.

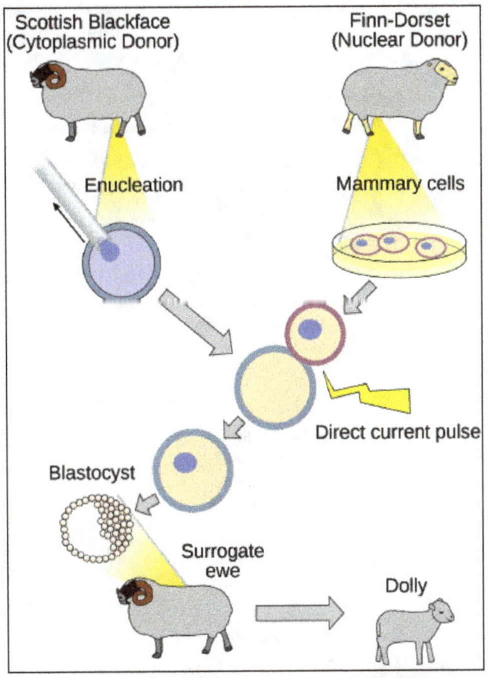

Figure shows dolly the sheep was the first agricultural animal to be cloned. To create Dolly, the nucleus was removed from a donor egg cell. The enucleated egg was placed next to the other cell, and then they were shocked to fuse. They were shocked again to start division. The cells were allowed to divide for several days until an early embryonic stage was reached, before being implanted in a surrogate mother.

The first cloned agricultural animal was Dolly, a sheep who was born in 1996. The success rate of reproductive cloning at the time was very low. Dolly lived for six years and died of a lung tumor. There was speculation that because the cell DNA that gave rise to Dolly came from an older individual, the age of the DNA may have affected her life expectancy. Since Dolly, several species of animals (such as horses, bulls, and goats) have been successfully cloned.

There have been attempts at producing cloned human embryos as sources of embryonic stem cells. In the procedure, the DNA from an adult human is introduced into a human egg cell, which is then stimulated to divide. The technology is similar to the technology that was used to produce Dolly, but the embryo is never implanted into a surrogate mother. The cells produced are called embryonic stem cells because they have the capacity to develop into many different kinds of cells, such as muscle or nerve cells. The stem cells could be used to research and ultimately provide therapeutic applications, such as replacing damaged tissues. The benefit of cloning in this instance is that the cells used to regenerate new tissues would be a perfect match to the donor of the original DNA. For example, a leukemia patient would not require a sibling with a tissue match for a bone-marrow transplant.

Steps in Gene Cloning

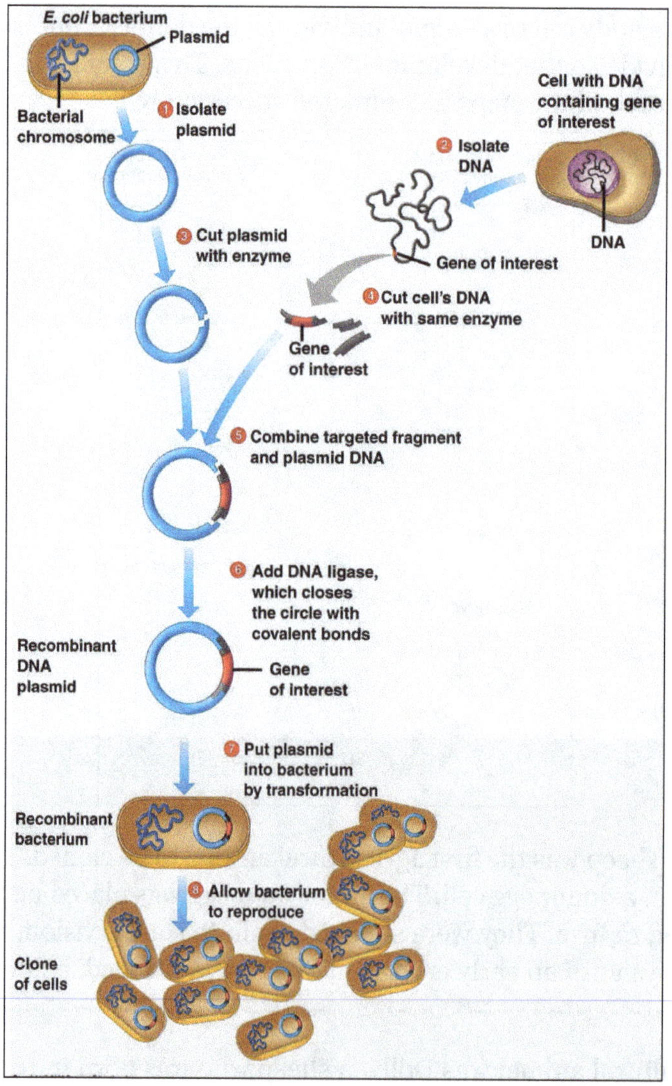

The basic 7 steps involved in gene cloning are:

- Isolation of DNA [gene of interest] fragments to be cloned.
- Insertion of isolated DNA into a suitable vector to form recombinant DNA.
- Introduction of recombinant DNA into a suitable organism known as host.
- Selection of transformed host cells and identification of the clone containing the gene of interest.
- Multiplication/Expression of the introduced Gene in the host.
- Isolation of multiple gene copies/Protein expressed by the gene.
- Purification of the isolated gene copy/protein.

Isolation of the DNA Fragment or Gene

The target DNA or gene to be cloned must be first isolated. A gene of interest is a fragment of gene whose product (a protein, enzyme or a hormone) interests us. For example, gene encoding for the hormone insulin. The desired gene may be isolated by using restriction endonuclease (RE) enzyme, which cut DNA at specific recognition nucleotide sequences known as restriction sites towards the inner region (hence endonuclease) producing blunt or sticky ends. Sometimes, reverse transcriptase enzyme may also be used which synthesizes complementary DNA strand of the desired gene using its mRNA.

Selection of Suitable Cloning Vector

The vector is a carrier molecule which can carry the gene of interest (GI) into a host, replicate there along with the GI making its multiple copies. The cloning vectors are limited to the size of insert that they can carry. Depending on the size and the application of the insert the suitable vector is selected. The different types of vectors available for cloning are plasmids, bacteriophages, bacterial artificial chromosomes (BACs), yeast artificial chromosomes (YACs) and mammalian artificial chromosomes (MACs). However, the most commonly used cloning vectors include plasmids and bacteriophages (phage λ) beside all the other available vectors.

Essential Characteristics of Cloning Vectors

All cloning vectors are carrier DNA molecules. These carrier molecules should have few common features in general such as:

- It must be self-replicating inside host cell.
- It must possess a unique restriction site for RE enzymes.
- Introduction of donor DNA fragment must not interfere with replication property of the vector.
- It must possess some marker gene such that it can be used for later identification of recombinant cell (usually an antibiotic resistance gene that is absent in the host cell).
- They should be easily isolated from host cell.

Formation of Recombinant DNA

The plasmid vector is cut open by the same RE enzyme used for isolation of donor DNA fragment. The mixture of donor DNA fragment and plasmid vector are mixed together. In the presence of DNA ligase, base pairing of donor DNA fragment and plasmid vector occurs. The resulting DNA molecule is a hybrid of two DNA molecules – the GI and the vector. In the terminology of genetics this intermixing of different DNA strands is called recombination. Hence, this new hybrid DNA molecule is also called a recombinant DNA molecule and the technology is referred to as the recombinant DNA technology.

Transformation of Recombinant Vector into Suitable Host

The recombinant vector is transformed into suitable host cell mostly, a bacterial cell. This is done either for one or both of the following reasons:

- To replicate the recombinant DNA molecule in order to get the multiple copies of the GI.
- To allow the expression of the GI such that it produces its needed protein product.
- Some bacteria are naturally transformable; they take up the recombinant vector automatically.

For example: *Bacillus*, *Haemophillus*, *Helicobacter pylori*, which are naturally competent.

- Some other bacteria, on the other hand require the incorporation by artificial methods such as Ca^{++} ion treatment, electroporation, etc.

Isolation of Recombinant Cells

The transformation process generates a mixed population of transformed and non-trans-formed host cells. The selection process involves filtering the transformed host cells only. For isolation of recombinant cell from non-recombinant cell, marker gene of plasmid vector is employed. For examples, PBR322 plasmid vector contains different marker gene (Ampicillin resistant gene and Tetracycline resistant gene. When pst1 RE is used it knocks out Ampicillin resistant gene from the plasmid, so that the recombinant cell becomes sensitive to Ampicillin.

Multiplication of Selected Host Cells

Once transformed host cells are separated by the screening process; becomes necessary to provide them optimum parameters to grow and multiply. In this step the transformed host cells are introduced into fresh culture media. At this stage the host cells divide and re-divide along with the replication of the recombinant DNA carried by them. If the aim is obtaining numerous copies of GI, then simply replication of the host cell is allowed. But for obtaining the product of interest, favourable conditions must be provided such that the GI in the vector expresses the product of interest.

Isolation and Purification of the Product

The next step involves isolation of the multiplied GI attached with the vector or of the protein encoded by it. This is followed by purification of the isolated gene copy/protein.

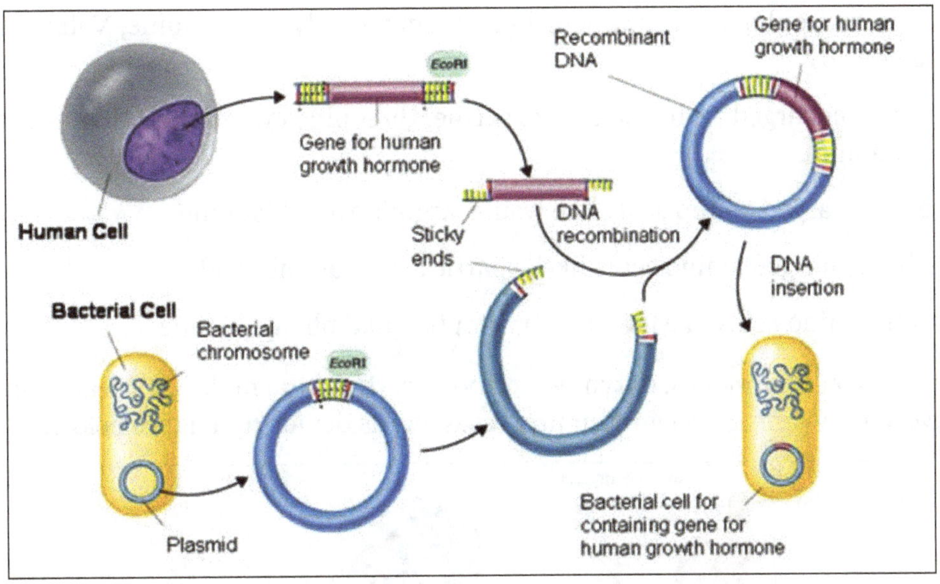

Applications of Gene Cloning

- A particular gene can be isolated and its nucleotide sequence determined.
- Control sequences of DNA can be identified & analysed.
- Protein/enzyme/RNA function can be investigated.
- Mutations can be identified, e.g. gene defects related to specific diseases Organisms can be 'engineered' for specific purposes, e.g. insulin production, insect resistance, etc.

Recombinant Proteins

In order to understand the collection/extraction of recombinant proteins and their subsequent purification, it is important to understand the structural components of a protein followed by the basic difference between naturally occurring proteins and recombinant proteins. Proteins are a group of biological macromolecules that are primarily involved in carrying out majority of functions inside/outside a cell, tissue or an organism as a whole; that is why proteins are referred to as "effector molecules". They are made up of 20 different types of naturally occurring amino acids. All amino acids (except glycine) contain a carbon atom (referred to as "alpha carbon") that is liked to four different groups, viz, carboxyl group, amino group, hydrogen atom and a side chain (R group) that is different in different amino acids. The presence of 4 different groups bestows chiral character to the alpha carbon. In contrast, glycine contains a carbon atom that is linked to a carboxyl group, an amino group and two hydrogen atoms. As the side chain of different amino acids is different, this imparts different reactivity to different amino acids that eventually is exploited during purification of recombinant proteins. Based on nature of R group and/or its ionization status in physiological conditions, amino acids can be classified as:

- Aliphatic, non-polar, whose side chain comprises of carbon chain. The longer the chain, the more hydrophobic the amino acid is. Example: Glycine, Alanine, Valine, Leucine and Isoleucine.

- Polar but uncharged amino acids like serine, threonine cysteine, methionine, asparagines and glutamine.
- Positively charged amino acids like lysine, arginine and histidine.
- Negatively charged amino acids like aspartic acid glutamic acid.
- Aromatic amino acids like tyrosine, tryptophan and phenylalanine.

Proteins majorly act as hormones, enzymes, antibodies, structural molecules, transport molecules, receptors involved in various recognition processes etc. as depicted in figure below.

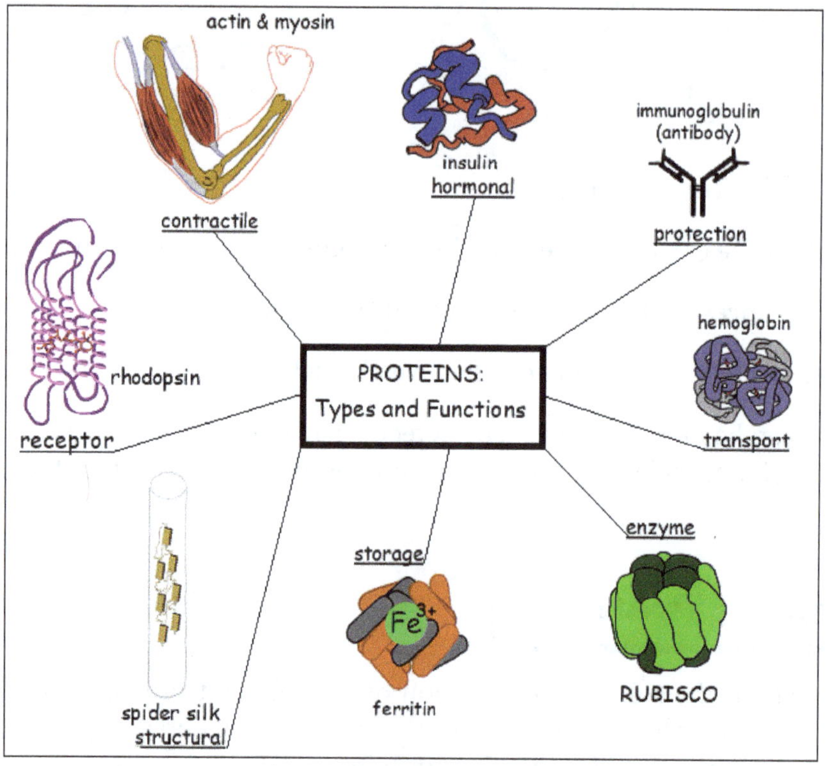

Figure: Functions of Proteins.

Recombinant proteins are those proteins that are synthesized in laboratory by inserting a foreign gene in host organism. This is achieved by first joining/ligating the gene of interest with the expression vector (for example plasmids in case of expression in bacterial system) through cloning. This expression vector is then inserted into the host cell (transformation) inside which the recombinant proteins are synthesized. The hosts that are normally used are bacteria (Prokaryotic host); yeast, mammalian cells, plant cells (Eukaryotic host). As many mammalian proteins need to be modified after synthesis (via post translational modifications like phosphorylation, glycosylation etc.), so for synthesis of such proteins, a eukaryotic host like yeast is used instead of bacteria as the bacterial system does not have required machinery for post translational modifications. The process of production of recombinant proteins in a bacterial production system is illustrated in figure.

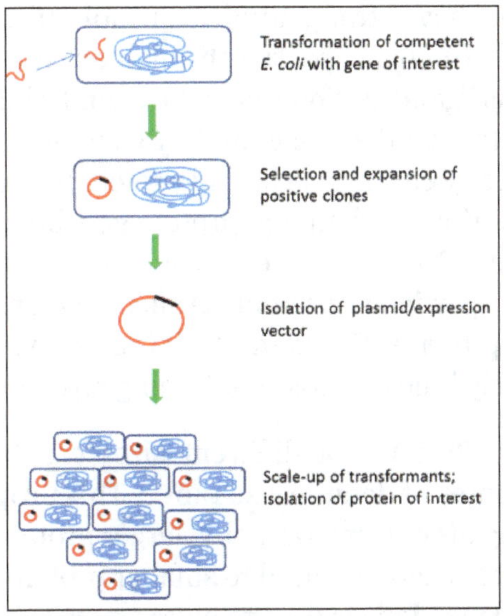

Figure: Bacterial protein expression system- Escherichia coli.

As proteins perform diverse physiological functions, they have immense applications in areas like human nutrition, therapeutics, cosmetics, animal husbandry etc. wherein they are needed in large quantities. Earlier, proteins used to be obtained by purification from biological sources but the processes were time consuming and yield from natural sources was very low. On the other hand, through recombinant DNA technology, large amount of desired proteins can synthesized in a short span of time. Some of the therapeutically important recombinant proteins include human insulin, interferon beta, human growth hormone, follicle-stimulating hormone, blood clotting factor VIII etc.

Purpose of Extraction and Purification of Recombinant Proteins

Every protein has a specific function. Before use, recombinant proteins need to be purified so that other proteins and various other biomolecules are removed. This removal is required as contaminating molecules may interfere with the function of recombinant protein and may have undesired effect. Once purified, further structural and/or functional characterization may be performed.

Besides, the production of high amount of recombinant proteins in bacteria also lead to the formation of inclusion bodies- a mass aggregated protein, that needs to be resolubilized before purification by using suitable solubilization buffer. Earlier high concentration of strong chaotropic agents like urea and guanidine hydrochloride were used to solubilize inclusion bodies, but since inclusion bodies largely consist of a single type of polypeptide with considerable secondary structure, buffers with low concentration of chaotropic agents have also been successfully used for solubilizing the inclusion bodies.

Properties of Proteins are Utilized for Protein Extraction

Following are certain properties of proteins that form the basis of separation of recombinant proteins by various purification methods:

- Charge: As the charge on proteins depends on the amino acid composition, different proteins have different net charge depending upon the pH of solution. This property is utilized for purification by techniques like ion exchange chromatography.

- Affinity: Certain proteins have strong affinity for molecules which are referred to as their ligands. If a mixture of proteins is passed through an immobilized ligand, the protein(s) which has affinity for this ligand gets bound to the ligand while other proteins can be washed out. The bound protein(s) can then be eluted by changing the ionic strength of the buffer or by changing the pH or by eluting the bound protein by an elution buffer containing high concentration of ligand. If more than one protein gets bound to the ligand then they may be purified in later stages. However, generally such a ligand is chosen which has strong affinity and is specific for a particular protein. In such cases, people have reported single step purification for various proteins. For example, if Concanavalin A (glucose binding lectin) is immobilized, it selectively binds glucose containing glycoproteins.
- Molecular Weight/Size: Proteins of different sizes can be separated by passing them through a matrix with pores of definite size range. This property is utilised in gel filtration chromatography. During his chromatography, larger proteins which do not enter the pores of matrix gets eluted out fast as compared to smaller proteins that move both via inside and outside of the pores and take longer time to elute.
- Hydrophobicity: The presence of hydrophobic amino acids, and therefore hydrophobic regions, on the surface to proteins is utilised to purify them by hydrophobic interaction chromatography.
- Stability: If the protein is stable at high temperature and over a range of pH, other heat labile and pH sensitive proteins can be removed by either heating the solution at temperature or adjusting the pH of solution such that the protein of interest retains its stability while other proteins gets denatured/aggregated out of solution.
- Solubility: Differential solubility of proteins in different solvents and solubility being affected by the presence of different cosolvents (example salts like ammonium sulphate), this property is routinely used in fractionating proteins before further purification by techniques like chromatography.

Extraction of Recombinant Proteins

After recombinant proteins are synthesized, they are either retained within the cell (Intracellular) or secreted in the periplasmic space/external medium (secretory proteins). Intracellular proteins may either be cytosolic or membrane bound proteins. As cytosolic and peripheral membrane proteins are largely hydrophilic they can be extracted using aqueous buffer. However, as integral membrane proteins have exposed regions mainly comprising of non-polar amino acids (that interact with hydrophobic lipids of membrane) their extraction and purification buffer needs to contain detergents (either ionic detergents like SDS, CTAB and CHAPS or nonionic detergents like Nonidet P-40, Tween-20 and Triton X-100).

Important Factors for Protein Extraction

Following are some important factors that need to be taken into account during extraction of recombinant protein:

1. Preparation of extraction buffer: For effective extraction, appropriate buffer with suitable

ionic strength and pH need to be prepared. Generally, 0.1-0.2M phosphate and Tris buffers are used in the pH range of 7.0-8.0. In addition certain other substances like antioxidants, substrate/enzyme inhibitors (in case of extraction of enzymes), cofactors, EDTA and polyvinylpyrrolidone are added to extraction buffer. As proteins inside are in a highly reducing environment in cell, so during extraction they are exposed to oxidizing environment. Therefore, in order to prevent oxidation antioxidants are added. During extraction of recombinant enzyme, an addition of substrate tends to stabilize the enzyme and facilitate the determination of its activity during purification. Presence of cofactors also performs similar function. When recombinant cells are lysed for extraction of intracellular proteins, the proteolytic enzymes (proteases) gets unpacked. So in order to minimise protein loss due to the action of proteases, extraction is carried out at low temperature (around 4°C) and in the presence of certain protease inhibitors.

Table: Protease inhibitors used in extraction buffers.

Sr. No	Class of protease inhibitors	Examples
1	Serine protease inhibitors	Chymostatin, 3,4-Dichloroisocoumarin, Di-isopropylphosphofluoridate, Phenylmethyl sulphonylfluoride, tosylphenylalanyl-chloromethylketone
2	Thiol protease inhibitors	acetylleucyl-leucyl-norleucinal, Iodoacetate, cystatin, geldanamycin, berkeleydione
3	Aspartic protease inhibitor	Pepstatin
4	Metalloprotease inhibitors	EDTA, 1,10-phenanthroline
5	Exopeptidase inhibitors	Amastatin, bestatin, puromycin, Dphenylalanine

When extraction buffer is required to be used for extended periods, antibacterial and/or antifungal agents are also added. Sodium azide is one commonly used bacteriostatic agent whose typical effective concentration usually ranges from 0.02% to 0.1% (w/v). Polyvinylpyrrolidone is generally used in case of extraction of recombinant proteins from plant tissues in order to prevent oxidation of phenolic compounds present in plant tissues. The oxidation products of these phenolic compounds cause cross-linking and aggregation of proteins, thereby interfering with the extraction process.

2. Methods for cell disruption: For extracting intracellular proteins recombinant cells need to be disrupted and for these blenders, grinders, presses, sonicators and various enzymatic methods are employed. Both blender and grinder (generally used with abrasives) use strong shear forces to break open the cells. In a press, cell suspension is passed through a small orifice under high pressure. As the cells emerge from orifice, the pressure gets reduced and the cells expand rapidly and get burst. During sonication, high frequency sound waves are used to break cells. As large amount of heat is evolved during sonication, the process is carried out using samples place in ice. The enzyme lysozyme is also used as it cleaves the thick peptidoglycan layer of gram positive bacteria. In case of gram negative bacteria like E. coli EDTA is used to remove calcium ions and thus lipopolysaccharide layer gets destabilized. Along with lysozyme, non-ionic detergents are also added to extraction buffer in order to permeabilise the cell membrane. In case of extraction of proteins from yeast, zymolyase and lyticase are commonly used while chitinase is used to disrupt filamentous fungi. Various cell disruption methods are depicted in figure.

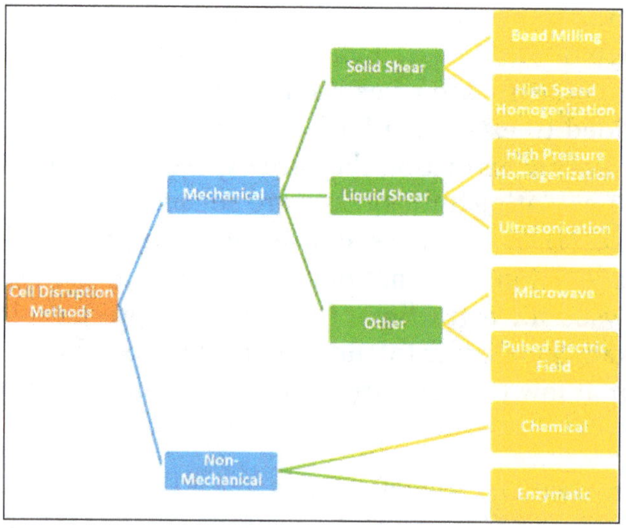

Figure: Classification of the cell disruption methods for releasing intracellular proteins.

Purification of Recombinant Proteins

After the protein of interest has been synthesized by the host cell, it needs to be purified for further characterization and/or use. As mentioned earlier, it is comparatively easier to purify secretory protein as compared to intracellular one. In either case, the protein fraction needs to be separated from cell debris/other cellular components. As the protein of interest may be hydrolyzed by intracellular proteases, protease inhibitors are often added to the purification buffer. Also, the purification of recombinant proteins is often carried out at low temperature, around 4°C, in order to minimise the activity of proteases. In case, the sample contains high amount of fat, the sample should be defatted by using highly non polar solvent like petroleum ether or hexane using soxhlett, percolation or racemization. It should be noted that during this process, the non-polar proteins would also be removed. Therefore, if the protein of interest is a membrane-bound protein (hydrophobic in nature), the defatting procedure may be avoided.

The next step in protein purification is the fractionation of proteins that is carried out using ammonium sulphate fractionation. This is due to 'salting out' effect that leads to the dissolved proteins getting removed from the solvent as the concentration of salt is gradually increased. If the recombinant protein is an enzyme, different protein fractions precipitated are then dissolved in buffer and they are analysed for the presence of desired protein by measuring the activity in different proteins. In case of non-enzymatic proteins, and if the antibody against the protein of interest is available then the fraction containing the protein of interest can be identified by techniques like enzyme linked immunosorbent assay (ELISA). Once the protein fraction containing the recombinant protein is identified, the protein can be further purified and identified by one or the combination of following techniques:

Chromatography

Principle

The separation during chromatography depends on how the molecule fractionates itself when it is passed through two immiscible phases, the mobile phase and the stationary phase. The ratio of the concentration of molecule in one phase divided by its concentration in second phase gives its distribution or partition coefficient, designated as K_D. So, if a mixture of compounds is applied onto a chromatographic column and mobile phase passed through it, different compounds have different mobility. The compound that distributes itself more into the mobile phase moves faster and gets eluted out earlier as compared to the compound that distributes itself more in the stationary phase. The recombinant protein can be separated by either one or the combination of following chromatographic separations:

- Gel Filtration Chromatography: The separation is based on the size or molecular weight of the protein. This method can also be used to determine the approximate molecular weight of the protein. Earlier, the separation used to be under gravitational force or low by using low pressure peristaltic pumps but nowadays high pressure pumps are used (in HPLC) that speed up the processing time and the separation is complete in minutes.
- Ion Exchange Chromatography: The separation is primarily based on the net charge on the protein at particular pH. Other factors like charge density, charge distribution and size of the molecule also play an important role.
- Hydrophobic Interaction Chromatography: The separation is accomplished by exploiting the interaction between hydrophobic regions on the surface of protein with the matrix and then eluting the bound protein by changing the polarity of solvent.
- Affinity Chromatography: This exploits strong affinity between the protein of interest and its ligand when a mixture of proteins is passed through an affinity column with immobilized ligand. Only the protein having affinity for ligand is bound while all other are removed during washing. The bound protein is then eluted by using a suitable elution buffer.
- Covalent Chromatography: While all the above methods involve non-covalent binding between the recombinant protein and the matrix, this involves covalent binding between protein and the matrix. The matrix used has bound molecules capable of forming disulfide bond with the free thiol groups of the protein. Recombinant proteins containing free thiol groups have been purified using this method.

Separation and Identification of Proteins by Electrophoresis

During electrophoresis, the protein molecules are made to move under the influence of electric field. Polyacrylamide gels provide the porous medium through which the proteins move towards their respective electrode with mobility that is determined by their net charge and/or size (molecular weight). Two commonly used types of electrophoresis are native PAGE (Polyacrylamide Gel Electrophoresis) and PAGE under denaturing conditions. The former being performed in the electrophoresis buffer that maintains the proteins in their native state while the latter is performed by denaturing the proteins in the presence of detergent, sodium dodecyl sulphates, that uniformly binds (On an average 1 SDS binds for every 2 amino acids) to protein and imparts negative charge. So, during native PAGE, the proteins are separated based on their mobility that is determined by their net charge and size while in SDS PAGE, the proteins are separated on the basis of their size. Once the proteins are separated they may transferred from gel to the nitrocellulose membrane (Western Blotting) on which the proteins are adsorbed. Once adsorbed, the protein of interest may be identified by incubating the nitrocellulose membrane with antibodies against the protein of interest. Another important application of electrophoresis is by performing electrophoresis in 2 dimensions (2D Electrophoresis), the first dimension wherein separation is based on isoelectric point (isoelectric focusing) and the second dimension during which the separation is on the basis of size (SDS PAGE). Using this approach, up to 3000-5000 different proteins can be easily separated from tissue extracts and proteins differing in just 1 amino acid can be separated. Proteins thus separated can be identified through western blotting.

Both electrophoresis and immunological methods (discussed in the following section) are useful for identification and characterization in case the recombinant protein is non-enzymatic while in case of enzymes their presence in different fractions may be ascertained during purification by measuring their activity.

Identification and Quantification of Proteins by Immunological Methods

If the antibodies against the recombinant protein are available then the presence of protein in different fractions during purification and its quantification can be done by enzyme linked immunosorbent assay (ELISA). During ELISA, the antibody (primary antibody that may either be monoclonal or polyclonal) against the protein of interest is immobilized on a suitable support and is then incubated with the solution/fraction that is to be screened. If the protein is present in the mixture, it would bind to the immobilized antibody. All non-specifically bound proteins are removed during washing. Addition of enzyme labeled secondary antibody (that is polyclonal if the primary antibody used is monoclonal or vice versa) leads to the formation of a sandwich of the bound recombinant protein between primary and enzyme labeled antibody (Sandwich ELISA). On addition of the substrate and its conversion to a colored product helps one identify as well as quantify the protein in the given fraction/mixture. For quantification a dose-response curve is first plotted between the amount of protein and intensity of color by taking different known concentrations of the protein. The color developed for the given fraction/mixture is then read on this curve and in this way, the amount of protein is determined.

Earlier, radiolabeled secondary antibodies were used in Immunoradiometric assay (IRMA), but the use of this technique has reduced due to the environmental impact of using the radioisotopes also due to the introduction of enzyme/fluorescent tag/Chemiluminiscent tag labeled secondary antibodies.

Strategies to Facilitate the Purification of Recombinant Protein

Proteins which are retained within the cells are difficult to purify as one needs to remove all other cellular proteins/ components. On the other hand, secretory proteins are comparatively easier to purify. That is why recombinant proteins are frequently expressed with signal peptide which is short peptide sequences at the N-terminal of protein that aid in either protein secretion or transporting it to compartments like periplasmic space that facilitate the extraction of recombinant protein for further purification. As bacteria does not have machinery for post translational modification of proteins, eukaryotic proteins which need post translational modification (like phosphorylation, glycosylation etc.) are expressed in eukaryotic host like yeast. As most such modifications are occur during the secretory pathway, ensuring the secretion of recombinant proteins also facilitate their modification if required. Besides, as one of major objectives of extraction is attaining high yield of production of recombinant proteins, ensuring the secretion of recombinant protein also prevents its degradation by intracellular hydrolytic enzymes. Another advantage of secretory proteins is that the host cells are prevented from toxicity as some of the recombinant proteins are toxic to the host cell, so their accumulation within the host cell is not desired.

Another method that facilitates the purification is to synthesize the recombinant proteins as fusion proteins, during which the recombinant protein is synthesized with another protein or tag. For fusion protein, the target gene is joined to another gene. This attached protein/tag is selected and/or designed in such a way so that it facilitates the purification and is easily removed either chemically of enzymatically after purification. One such example is the fusion of glutathione transferase gene with the protein of interest. After synthesis, the fusion protein can be purified by affinity chromatography using an immobilized glutathione. After purification, the fusion protein is cleaved by an enzyme that identifies specific sequence in the linker region. Another example of the fusion protein is attaching the gene for poly(arginine) with the gene of protein of interest. After synthesis, the fusion protein that is highly positively charged due to arginine tail at its Cterminal, can be purified by cation-exchange chromatography. The poly(arginine) tail is subsequently removed by Carboxypeptidase B that removes arginine from C-terminal of a protein.

Cell and Tissue Engineering

Tissue Engineering (TE) has emerged as essential component of regenerative medicine. This approach has high potential to address major problems we are facing with organ transplantation. TE is an interdisciplinary field which is based on bioengineering, material and biological sciences. By applying principles and methods of these fields we can aggregate biological substitutes to improve and restore normal functions of damaged tissue results drum disease condition. TE also allows us to have insight of human physiology.

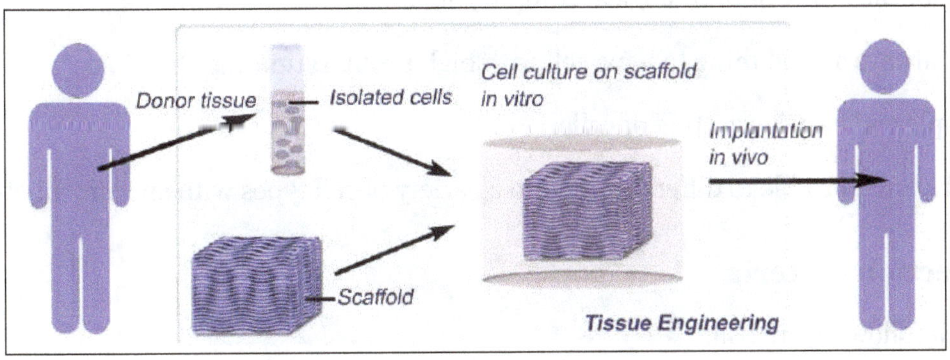

Figure: Principle of tissue engineering.

Cells in our body grow within an organized 3D structure known as extra cellular matrix (ECM) which decides whether a given cell will undergo proliferation, differentiation, invasion and apoptosis. The principle of TE involve growing a cell on natural/biodegradable synthetic support known as scaffold that build 3D living construct that look similar and work in same way or better than tissue to be replaced.

Approaches to Tissue Engineering

Approaches for TE can be categorized as below:

- Substitutive: It is an ex vivo approach where whole organ get replaced.

- Histioconductive: it is also an ex vivo approach but here just part of damaged tissue is replaced with normal constructs.
- Histioinductive: it may involve gene therapy or facilitate self-repair. In case of gene therapy The DNA is delivered using plasmind vector.

Requirements of Tissue Engineering

To develop tissue construct a careful selection of following materials is required:

- Cells
- Scaffolds
- Growth factor
- Extracellular matrix (ECM)

Cells

Natural as well as artificial tissues composition includes various cellular elements and their associated ECM. The interaction of cells with ECM occurs due to multicellular networks. For production of artificial tissue, it becomes necessary to have a comprehensive assessment of cells and tissue environment. An appropriate cell supply is required for viable construct of tissue. A cell must have following features:

- Do not cause any immunological complications.
- It is always a good thing to have cell with high proliferation rate.
- Simple and easy harvesting of cells.
- Cell should be able to differentiate into a variety of cell types with specific functions.

Cells Selection Criteria

It must fulfil following criteria:

- Quantity: An appropriate number of cells must be present at the site of defect.
- Differentiation: Cells differentiation is must to get desired phenotypes.
- Adaptation: Adaptation of cells must to 3D structural support i.e. scaffold is essential requirement and only adapted cell will produce ECM.
- Structure: Cells produced by ECM must possess structural and mechanical similarities with the native cell.
- Integration: Cells must integrate itself with native cells and minimizes the risk of immunological rejection.

Source of Cells

Various types of cell sources have been used for utilized in tissue engineering. These have been categorized on the basis of their isolation source as:

- Autologous: These are isolated from the patient. They are used as potential source of cells for TE because of minimum immunological problems however they pose high cost when used in clinical trials.
- Allogeneic: Cells are isolated from immunologically non-identical human donor. They offer advantages such as uniformity, quality control and cost effectiveness.
- Xenogenic: These are isolated from a different species donor such as animal cells in human patient.
- Syngenic/Isogenic: Isolated from genetically identical donor such as homozygous twin.

Cell sources can be further divided into non-stem (mature) and stem (adult, somatic, embryonic and totipotent) cells. The utilization of non-term cells is limited because of their low proliferation and differentiation power. Adult stem cell (ASC) have been found to reside in specific niches or tissue and plays important role in maintaining the integrity of tissues like skin, bone and blood. These are undifferentiated cells but can differentiate into specific cell type when they are programmed. At least 20 major categories of ASC have been recognized in mammals. They are present in tissue like bone marrow, blood, cornea and retina of the eye, dental pulp, liver, skin, GI tract, and pancreas.

Embryonic stem cells (ESCs) are isolated from inner cell mass (ICM) of the pre-implanted blastocyst. They are immature and undifferentiated cells but have ability to differentiate and possess capacity of unlimited self-renewal. They are believed to engage in development all three embryonic germ layers (ectoderm, mesoderm and endoderm). Both ASCs and ESCs are being researched for their potential to differentiate into the specific cells types for the purpose of repair of damaged bone, cartilage, and tendon. The development of human ESC lines and their ability to differentiate to cardiomyocytes is key achievement. Kehat et. al. first suggested the role of human ESCs in generating a reproducible spontaneous cardiomyocyte differentiation system. Several stem cells types such as adult hematopoietic (HSCs) and mesenchymal stem cells (MSCs), have been found to differentiate into hepatocytes under the appropriate environment. Ma et. al. demonstrated formation of neural network and functional synapse from central nervous system (CNS) derived stem cell on a 3D collagen gel. Totipotent stem cells or zygotes have the potential to develop into any cell found in the human body. They can replicate infinite times without any loss of total potency. The stem cells either (ESC) or adult pluripotent cells that are able to differentiate themselves into any cell type but they have restricted use due to ethical implications. Yamanaka and collaborator generated induced Pluripotent Stem Cells (iPSCs) is under consideration for their application in TE. iPSCs can be reprogrammed to a pluripotent state by introduction of certain set of transcription factors. The use iPSC is advantageous because of differentiation capacity, robustness and simplicity of reprogramming procedure. However, there are several barriers needs to be overcome before making extensive use of iPSCs.

Scaffolds

It provides framework of three-dimensional matrix that allows differentiation growth and organization of cells and tissue by production of an extracellular matrix (ECM).They are biocompatible and designed in such a way that they should fulfil nutritional as well biological needs of specific cell population. Scaffolds are supposed to perform few or all of following functions: (i) it increases cell-to-cell communication and adhesion (ii) It also favour ECM production & deposition ECM (iii) It helps in gaseous exchange, transport of nutrients and regulation of factors that controls the cell differentiation and proliferation (iv) Control rate for biodegradation is almost same as needed for tissue regeneration (v) In vivo risk of toxicity and inflammation is very minimal.

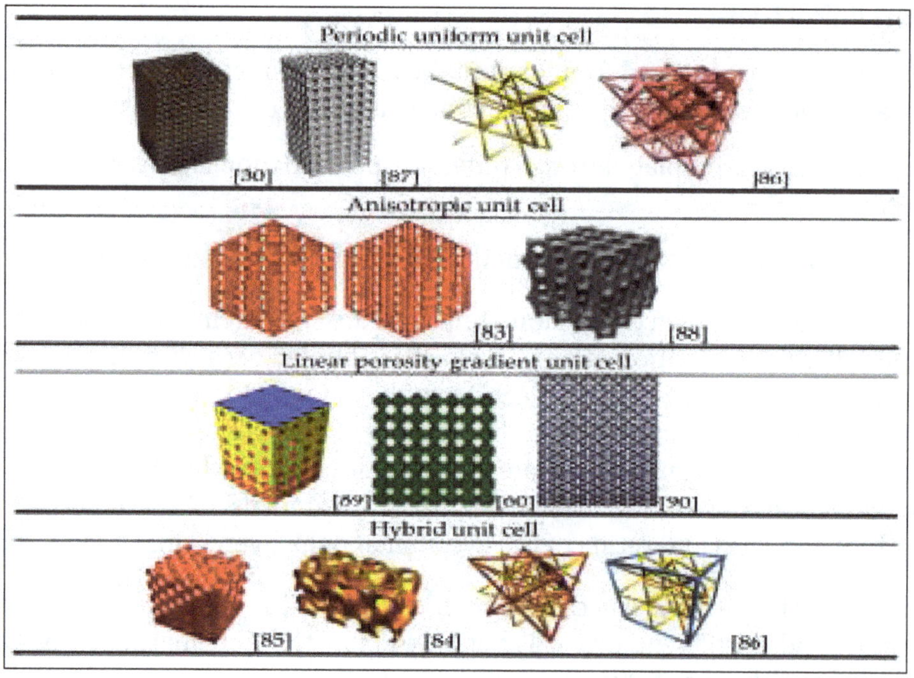

Types of scaffold.

In tissue engineering, various types of scaffolds used, on which seeding of cells is done. These can be categorized depending on its origin of material:

- Natural biomaterials: A main source for such biomaterials includes fibrin, collagen, gelatin, chitosan, alginate, hyaluronic acid etc. However, composition of this biomaterial depends on animal origin and isolation and purification procedures. They can be accessed with ease and in vivo they provide wide range of signals to different cell types.
- Synthetic polymers: They are commonly used in tissue engineering because of their porosity, and less degradation time. These properties make it highly effective for specific applications. These are often cheaper and possess long shelf time than biological scaffolds. They can produce in large quantities. They show physicochemical and mechanical properties similar to those of biological tissues. They belongs to a group of biodegradable polymers such as polyglycolic acid (PGA), poly(d, l-lactic-co-glycolic acid) (PLGA), polycaprolactone (PCL), polylactic acid (PLA), carbon nanotubes (CNT), polyvinyl alcohol (PVA), polyethyleneglycol (PEG), polyurethanes, and TiO_2 nanotubes.
- Hybrids materials: natural biomaterial provides biocompatibility while, synthetic biomaterial acquires tuneable mechanical properties. They are biodegradable and consisting of both synthetic polymer (PLGA) and naturally derived polymer (collagen). They are used as 3D scaffold for TE of bovine articular cartilage.

Extracellular Matrix (ECM)

The ECM represents heterogeneous connective tissue "network". It primary role is to provide physical support to growing cells and tissues. In nature, chondroitin and heparin sulphate couple with glycoproteins and stabilize the tissue mechanics. The ECM highly porous nanostructure provides them three important functions: (i) provide binding site for various types of soluble growth factors (GF), enzymes and other effector molecules and control their diffusion. As a result, ECM regulates gene expression and participates in cell phenotype by integrating itself in intracellular signalling pathways. They also results in mechanical signalling.

Growth Factors

The term growth factor is generally refers to naturally occurring substances that can stimulate differentiation and proliferation of a cell. The GF are critical for the maintenance of tissue integrity and cell to cell communication, their activities. Based on the critical role of growth factors in tissue development and repair, these multifunctional peptides should be involved in the tissue engineering strategies. GF are the necessary components to stimulate the correct biological signals for tissue integration and growth. Transforming growth factor (TGF), fibroblast growth factors (FGF) and insulin-like growth factor (IGF) are primary GF involved in chondrogenic differentiation. The TGF-β super-family includes many proteins which play essential role in the cartilage formation. Proliferation, differentiation, and production of articular cartilage and growth plate chondrocytes is regulated by FGF family. In IGF family, IGF-1 mediates chondrogenesis by increasing the production of proteoglycan and collagen type II.

Methodolgy

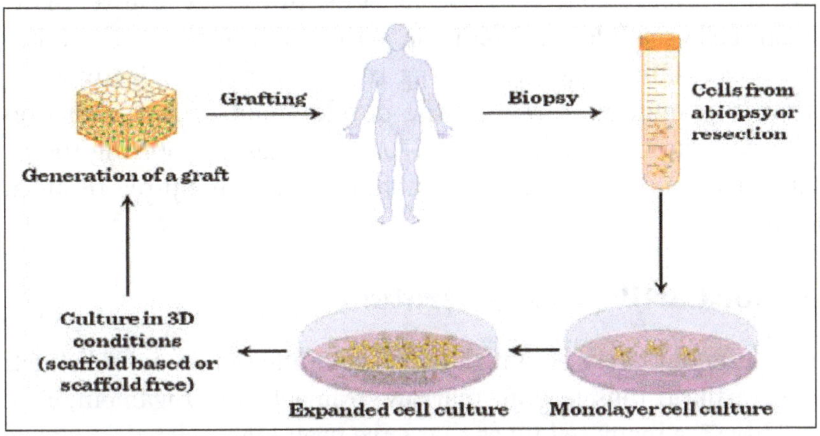

Figure: Process of tissue engineering.

Cells are building block of body and control all the activities. They support themselves by secreting their own support structures known as extra-cellular matrix (ECM). Other than support, ECM also have role in signalling pathway. With the help of ECM, cells receive messages in form of a signal from different sources and response of each signal predicts what will happen to cell. By understanding the mechanism of how each cell respond to signals, organize into tissues and interact with environment, scientist has been able to modify biological processes to repair damaged tissues or even create new ones. The process of TE can be divided into following steps:

- Extraction of damage tissue from donor by biopsy.

- Isolate desired cell type.
- Cells are grown under in vitro condition.
- Generate graft by expanding the cell culture on 3-D support (scaffold) by providing necessary components like growth factors and ECM.
- Transfer the graft having repaired tissue back to the donor.

Cartilage Tissue Engineering

Cartilage is one of the critical tissues existed in human and animal bodies, such as rib cage, ear, nose, bronchial tubes, intervertebral discs, meniscus, and the joints between bones. Cartilage injuries are the most common diseases. According to National Health Interview Survey (NHIS), in 2010–2012, about 52.5 million adults in the USA had doctor-diagnosed arthritis, and by 2040, the number of US adults with doctor-diagnosed arthritis is projected to increase 49% to 78.4 million. That means about 25.9% of all adults have arthritis. Degeneration of the intervertebral disc, a fibrocartilaginous joint residing between adjacent vertebrae in the vertebral column, is the most frequent cause of low back pain and another significant cartilage-related disease. The overall cost of chronic low back pain exceeds the combined costs of stroke, respiratory infection, diabetes, coronary artery disease, and rheumatoid disease. However, the damaged cartilage has little ability for repairing itself due to the lack of blood supply, nerves, and lymphangion, and the effective therapeutic treatments for cartilage regeneration are very few.

Tissue engineering is an interdisciplinary field that applies the principles of engineering and the life sciences toward the development of biological substitutes that restore, maintain, or improve tissue function. Stem cells scaffold, and biologic active molecules are three key components in tissue engineering. Successful tissue engineering relies on multiple factors including obtaining appropriate cells for implantation, directing the development of those cells on an appropriate differentiated pathway using growth factors and/or cytokines, supporting the growing cells on a three-dimensional matrix, and having that matrix remains in the injured tissue area, at least until healing is completed.

Stem Cells used for Cartilage Regeneration

Stem cells have multi-differentiation potential, which can differentiate into distinctive end-stage cell types including bone, cartilage, muscle, bone marrow stroma, tendon/ligament, fat, dermis, and other connective tissues. There are many cell types that have been manipulated *in vitro* and subsequently implanted to repopulate a cartilage defect. It must be ensured that the implanted cells are immunoprivileged or provide immunosuppressive agents to avoid rejection by the host immune system.

Autologous Chondrocytes

Autologous chondrocytes were first used for the treatment of cartilage defects of the patients by a Swedish group in 1994. This approach needs a slice of healthy articular cartilage obtained arthroscopically from proximal part of the medical femoral condyle of the affected knee joint during the first operation. The chondrocytes were isolated from this healthy articular cartilage and cultured for 2–3 weeks to prepare enough cells (about 5×10^6) for damaged cartilage repair. The clinical studies have shown that the treatment of autologous chondrocytes prompts pain reduction, improves quality of life, and delays the need of joint replacement in many cases. Despite the encouraging clinical results, there are still limitations to the use of autologous chondrocyte transplantation. The conventional technique is accompanied with periosteum harvest and fixation over the cartilage defects via large skin incisions. Autologous chondrocytes were injected underneath the periosteal flap. Hypertrophy of the periosteum with high rate of revision arthroscopies and the risk of transplant failure of up to 20% are major drawbacks of the conventional autologous chondrocyte transplantation. Moreover, the complexity and the cost of the two surgical procedures, the biological response of the periosteal flap, and the de-differentiation and consequent capacity loss associated with *in vitro* expansion of isolated chondrocytes are also the limitations.

Bone Marrow-derived Mesenchymal Stem Cells (BMSCs)

Mesenchymal stem cells (MSCs) are multipotent stromal cells first identified and described in 1966 by Alexander Fridenstein. Adult MSCs were originally isolated from bone marrow in 1999 by Pittenger and his colleagues. Subsequent studies have demonstrated that MSCs present in various parts of the body including bone marrow (BM), peripheral blood, umbilical cord blood, fatty tissues, skeletal and cardiac muscles, Wharton's Jelly of umbilical cord, facet joints, interspinous ligaments, and ligamentum flavum. Many studies have shown that MSCs can migrate to injury sites, induce peripheral tolerance, and inhibit the release of proinflammatory cytokines. It has been demonstrated that MSCs can also promote tissue repair and survival of damaged cells. However, it is not clear which adult tissue-derived MSCs should be used as a good source for cartilage repair.

Autologous bone marrow mesenchymal stem cell (BMSCs) transplantation was first used for the repair of full-thickness articular cartilage defects in human patellae by a Japanese group. BMSCs were aspirated from iliac crest and the nucleated cells were cultured. Adherent cells were subsequently collected, embedded in a collagen gel, transplanted into the articular cartilage defect in patellae, and covered with autologous periosteum. Six months after transplantation, clinical symptoms (pain and walking disability) were improved and the improvement was persisted for 9 years post-transplantation. Sixteen years after transplantation, no clinical problem has been reported. Human autologous BMSCs have been used successfully to treat articular cartilage defects. Twelve months after BMSC transplantation, magnetic resonance imaging (MRI) revealed complete defect fill and complete surface congruity with native cartilage. Currently, autologous BMSC transplantation has been widely used for cartilage repair. Although BMSC treatment did not require any cell expansion or manipulation, reducing costs, and risks involved, the quantity of bone marrow cells was somewhat unsatisfactory.

Adipose-derived Stem Cells (ADSCs)

Among MSCs, adipose-derived stem cells (ADSCs) have been recognized as an appropriate cell type with chondrogenic potential and high proliferation capacity. Approximately 400,000 liposuction surgeries are performed in the USA each year, and these procedures yield anywhere from 100 ml to 3 liters of lipoaspirate tissue. This material is routinely discarded. It is well known that adipocytes are developed from mesenchymal cells via a complex cascade of transcriptional and non-transcriptional events that occur throughout the human life. Thus, adipose tissue is a good stem cell source.

The initial methods to isolate cells from adipose tissue were developed by Rodbell and colleagues.

They isolated adipose-derived stromal cells from rat fat pads by four steps. Step 1: Adipose tissue was minced into small pieces. Step 2: The adipose tissue pieces were digested with collagenase. Step 3: The cell pellet was obtained by centrifuge. Step 4: The cell pellet was cultured for future use. This protocol has been widely used for the isolation of adipose-derived stem cells (ADSCs) from human adipose tissues with some modifications.

The adipose tissue can be collected by needle biopsy or liposuction aspiration. The collected adipose tissues should be washed with 5% penicillin/streptomycin (P/S)-containing phosphate-buffered saline (PBS) twice, and then the tissue samples should be put in a sterile tissue culture plate and cut into small pieces. The minced tissues are digested with 0.075% collagenase at 37°C for 30 min; the collagenase is removed by centrifuging the digested solution (adipose tissue and collagenase mixture) at 1200 g for 10 min; the adipose-derived stem cells-containing pellet is then resuspended with culture medium (alpha-MEM, Mediatech, Herndon, VA) supplemented with 20% of fetal bovine serum (FBS), 1% L-glutamine (Mediatech, Herndon, VA), and 1% penicillin/streptomycin (Mediatech, Herndon, VA). The cell suspension is filtered through 70-μm cell strainer and cultured in a humidified tissue culture incubator at 37°C with 5% CO_2. The medium is changed every second day until the cells reach 80–90% confluence. It is important that the adipose tissue should be treated within 24 hours, and the cells isolated from about 500 mg of adipose tissue should be added into one well of 12-well plates.

Adipose-derived stem cells (ADSCs) are readily accessible with no morbidity and display the capability to differentiate into several cell lineages, including the spontaneous chondrogenic differentiation. Compared with bone marrow-derived MSCs, adipose-derived MSCs from lipoaspirates are acquired using a less invasive procedure and are in large amounts. ADSCs have been used for the repair of articular cartilage defect in non-weight-bearing areas.

Synovial-derived Stem Cells (SDSCs)

Synovial-derived MSCs have been isolated from human synovial fluid and synovium of the knee and the hip using the following protocols. The synovial tissue samples (wet weight 10–50 mg) were obtained aseptically from the joints and rinsed twice with Hanks' balanced salt solution (HBSS; Life Technologies, Carlsbad, CA) supplemented with antibiotic-antimycotic solution (100 units/ml penicillin, 100 μg/ml streptomycin, and 0.25 μg/ml amphotericin B; Life Technologies; Carlsbad, CA). The washed tissues were minced into small pieces and digested with 0.5 ml of 0.2% collagenase (Life Technologies, Carlsbad, CA) in high-glucose Dulbecco's Modified Eagle's Medium (DMEM; Life Technologies, Carlsbad, CA) at 37°C for 1 hour. The digested solution were removed by centrifugation at 1500 g for 10 min; the SDSCs-containing pellet was resuspended in growth medium (high-glucose DMEM supplemented with 10% FBS and 100 units/ml penicillin, 100 μg/ml streptomycin) and cultured in a humidified tissue culture incubator at 37°C with 5% CO_2. The medium was first changed at day 7 and changed every 3 days until the cells reach 80–90% confluence. It is important that the synovial tissue should be treated within 24 hours.

SDSCs obtained by above procedures have a higher proliferative capacity and chondrogenic potential than the MSCs derived from other sources. A small synovial tissue biopsy is an easily accessible source of autologous MSCs in the context of an explorative or therapeutic arthroscopy. These cells can be subsequently used for the regeneration of damaged cartilage. Autologous chondrocyte transplantation used for cartilage defect repair is limited by the availability of cells, particularly in elderly individuals, and by the well-known dedifferentiation events associated with chondrocyte expansion. Furthermore, SDSCs can be harvested relatively in a minimally invasive manner from synovial fluid and retain a particularly high capacity for chondrogenic differentiation and proliferation compared with MSCs obtained from other tissues, such as bone marrow or cartilage, those have second injury on healthy tissues. SDSCs may be an optimal alternative source of chondrogenic cells for cartilage defect repair.

A recent research has shown that xenogenic implantation of equine SDSCs into rat cartilage defect area leads to articular cartilage regeneration. Horse joints are anatomically equivalent to the human knee and ankle; as a result, horses are widely used as large animal preclinical models for cartilage repair studies. However, large animal studies pose logistical and financial challenges, and small animal rodent models are cost-effective and have proven to be useful for proof-of-concept studies. There was no any immune response to the equine cells in the treated rat knees. This result was also confirmed by a xenogenic transplantation of human MSCs in a critical size defect of the sheep tibia for bone regeneration. Another xenogenic transplantation study has shown that human MSCs can enhance damaged pig intervertebral disc regeneration.

Growth Factors used for Cartilage Regeneration

Growth factors play an important role in cartilage regeneration.

Transforming Growth Factor-beta Family (TGF-β)

In cartilage repair, the four most thoroughly investigated members of TGF-β superfamily are TGF-β1, TGF-β3, bone morphogenetic protein-2 (BMP-2), and bone morphogenetic protein-7 (BMP-7). It has been reported that TGF-beta 1 stimulates chondrocyte synthetic activity and decreases the catabolic activity of IL-1. TGF-β3 has been used to simulate extracellular matrix (ECM) synthesis in rabbit cartilage injury. Bone morphogenetic proteins (BMPs) play an important role in the development of bone and cartilage. They are involved in the hedgehog pathway, TGF beta signaling pathway, and in cytokine-cytokine receptor interaction. Animal studies have shown that BMP-2 enhanced cartilage matrix production and blocked the IL-1-induced cartilage degeneration. BMP-7 is another gold standard growth factor for cartilage repair. It has been reported that BMP-7 inhibits cell proliferation but stimulates ECM synthesis in both SDSCs and BMSCs.

Insulin-like Growth Factor-I (IGF-I)

IGF-1 is a multifunctional growth factor. The studies have found that IGF-1 play an important role in maintaining articular cartilage integrity. IGF-I deficiency has led to the development of articular lesions. IGF-1 can not only enhance the synthesis of proteoglycans and upregulate the gene expression of collagen II but also can reduce the degradation of extracellular matrixes by decreasing the production of matrix metallopetidase-13 (MMP-13). The research has shown that IGF-1 exerts these functions in a dose-dependent manner. Low dose of IGF-1 has a beneficial effect on bone remodeling by increasing bone formation markers in serum. Higher IGF-1 levels in osteoarthritis (OA) osteoblasts could be correlated with bone sclerosis.

Fibroblast Growth Factor (FGF)

There are two FGF members used in cartilage regeneration. One of them is called as basic FGF (bFGF) or FGF-2, and the other one is called as FGF-18. FGF-2 increases anabolic material levels and decreases aggrecanase levels in cartilage. In vivo study has indicated that bFGF can promote cartilage repair. However, some study indicated that the concentration of FGF-2 in synovial fluid samples of OA patients is approximately twice that of normal healthy knee joints. Further studies found that FGF-2 promoted the repair of partial thickness defects of articular cartilage in immature rabbits but not in mature rabbits.

A rat model study has shown that FGF-18 stimulates chondrogenesis and cartilage repair in a concentration-dependent manner. More studies have demonstrated that FGF-18 may present a therapeutic agent for osteoarthritis. A recombinant form of human FGF-18 has been used for cartilage injury treatment.

Platelet-rich Plasma (PRP)

Platelets play a fundamental role in hemostasis and are a natural source of growth factors. More than 30 growth factors have been identified in PRP; among them, the following six growth factors play an important role in cartilage regeneration. They are TGF-β1, platelet-derived growth factor (PDGF), fibroblast growth factor (FGF), insulin-like growth factor 1 (IGF-1), epidermal growth factor (EGF), and vascular endothelial growth factor (VEGF).

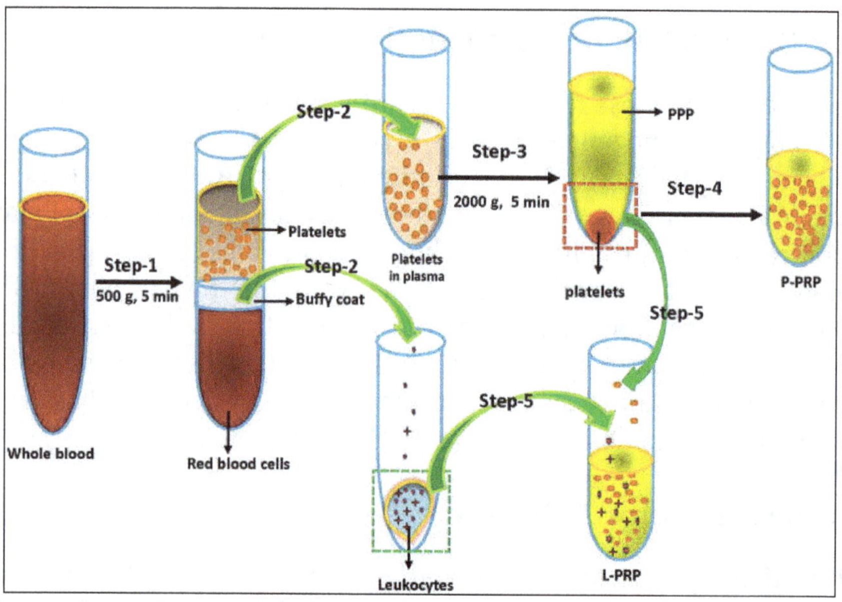

Figure: Scheme of preparation of P-PRP and L-PRP from whole blood using five steps.

The concentration of platelet in PRP used for cartilage repair should be two to three times higher than that of baseline. PRP can be prepared by the following five procedures. Step 1: blood (9 parts) is added into 3.8% sodium citrate solution (1 part) in a centrifuge tube and centrifuged at 500 g for 5 min to obtain three layers. Step 2: The supernatant at the top layer is transferred into a new tube, which is called as platelets-containing plasma, and the middle layer is transferred into another new tube, which is called leukocytes-containing plasma. Step 3: The platelets-containing plasma is centrifuged at 2000 g for 5 min to separate platelet-poor plasma (PPP) from the platelet pellet. Step 4: The platelet pellet is resuspended with appropriate amount of PPP to make pure PRP (P-PRP). Step 5: The leukocytes-containing plasma is mixed with platelet pellet and resuspended with appropriate amount of PPP to make leukocytes-containing PRP (L-PRP). Both P-PRP and L-PRP can be used for cartilage tissue engineering.

Bioactive Molecules used for Cartilage Tissue Engineering

Bioactive molecules used in cartilage tissue engineering include two kinds of materials: one is small molecular weight bioactive compound and the other one is high molecular weight materials including some nature biomaterials and synthetic polymers. Both of them play critical role in cartilage tissue engineering.

Kartogenin (KGN)

Kartogenin (KGN), a small heterocyclic molecule, has been discovered to enhance chondrogenic differentiation of human MSCs by regulating the CBFbeta-RUNX1 transcriptional program. Animal studies have shown that KGN can promote rabbit meniscus regeneration and wounded rat enthesis repair. *In vitro* and *ex vivo* experiments showed that KGN can reduce nucleus pulposus cell degeneration induced by interleukin-1beta (IL-1β) and tumor necrosis factor-alpha. More recent studies indicated that KGN inhibited pain behavior, chondrocyte inflammation, and attenuated osteoarthritis progression in mice; enhanced collagen organization and mechanical strength of the repaired enthesis of mouse rotator cuff; and induced chondrogenic differentiation of dental pulp stem cells.

These findings invigorate research into small-molecule therapy and regenerative medicine for cartilage diseases. It also provides new insights into the control of chondrogenesis that may ultimately lead to a stem cell-based therapy for osteoarthritis (OA). KGN and other structurally related small molecules that can promote selective differentiation of MSCs into chondrocytes may prove to be extremely useful for improving the outcome of cell-based therapy by stimulating endogenous mechanisms for repair of damaged cartilage, thus enhancing the joint's intrinsic capacity for cartilage regeneration.

Simvastatin

Simvastatin is a kind of HMG-CoA reductase inhibitor, which is widely used therapeutically to reduce morbidity and mortality in patients with hyperlipidemic cardiovascular disease. In addition to lowing low-density lipoprotein (LDL) cholesterol, statins have broad-range pleiotropic effects, including anti-inflammatory effects, which could exert an effect on synovium and cartilage. Animal studies found that simvastatin markedly inhibited not only developing but also established collagen-induced arthritis. Simvastatin inhibited the IL-6 and TNF-α production of human chondrocytes and cartilage explants in a concentration-dependent manner. Higher concentrations of simvastatin decreased nitric oxide (NO) production in both of human chondrocytes and cartilage explants. Statin treatment has also been shown to positively regulate components of the extracellular matrix in a rabbit OA model. More studies have shown that local application of simvastatin enhanced tendon-bone interface healing in rabbits. These studies have shown that the effect of simvastatin on articular chondrocytes may provide novel insight regarding the role of cholesterol homeostasis and signaling during cartilage development.

Biomaterial Scaffolds for Cartilage Tissue Engineering

Biomaterial scaffolds play an important role in cartilage tissue engineering, which act as a carrier to deliver the cells and bioactive molecules to the damaged tissue areas and also work as a template for tissue regeneration, to guide the growth of new tissue. There are two groups of biomaterial scaffolds used for cartilage tissue engineering. They are synthetic polymers and natural polymers. Commonly used natural materials in cartilage repair are agarose, alginate, chitosan, collagen, fibrin, and hyaluronan.

Agarose is a galactose polymer, which is suitable for cell encapsulation, especially for chondrocytes. When the ADSCs were cultured in agarose, they were differentiated into chondrocytes as evidenced by upregulation of the production of glycosaminoglycan (GAG). Moreover, dynamically loaded cell-seeded agarose hydrogel provided better graft tissues in a repair model of full thickness defects in rabbit joint cartilage. PRP combined with agarose as a bioactive scaffold has shown to enhance cartilage repair.

Another extensively studied natural scaffold used for cartilage tissue engineering is alginate, which is a polysaccharide extracted from brown algae. Generally, alginate is hydrophilic and water-soluble, thickening in neutral conditions, which is of great importance for *in situ* hydrogel formation. The good gelling properties of alginate-based scaffolds allowed them to be used as an injectable scaffold for the damaged cartilage repair. Human dental pulp stem cells were cultured in 3% alginate hydrogel and implanted in a rabbit damaged cartilage area. Three months after surgery, significant cartilage regeneration was observed. More studies have been done by mixing the cells or/ and growth factors with alginate solution to form gel microspheres in an isotonic $CaCl_2$ solution. The findings have shown that the cells are distributed homogeneously inside the gel microspheres. Those cell-containing alginate beads can be used as chondrogenesis-promoting scaffolds for cartilage regeneration.

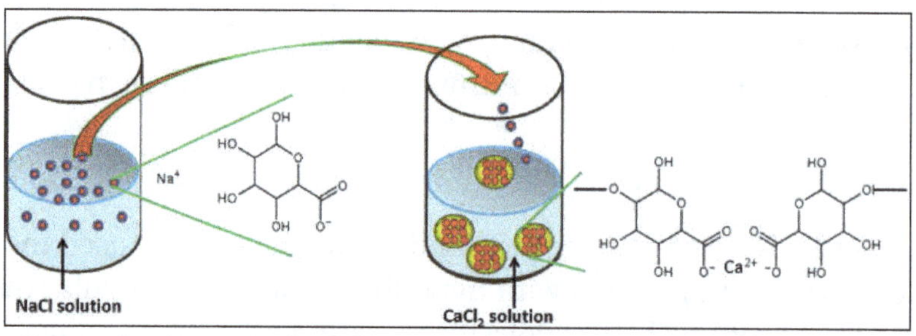

Figure: The intermolecular network of alginate molecules is formed in calcium chloride solution. Alginate can be dissolved with sodium chloride (left image), but cross-linked each other in calcium ions-containing solution to form hydrogel (right image).

Chitosan is another natural polysaccharide extracted from crustacean shells, particularly from shrimps and crabs. Chitosan contains glucosamine and hyaluronic acid (HA), which are basic components of the native cartilage. Therefore, chitosan is widely used for cartilage tissue engineering. The recent studies have shown that chitosan-hyaluronic acid hydrogel promoted wounded cartilage healing in a rabbit model.

Collagen is a main component of the extracellular matrix (ECM) of chondrocytes. Collagen gel has been widely used as substrates for articular cartilage substitutes. Injectable type II collagen gel has been used to treat full-thickness articular cartilage defects. Clinical study has demonstrated that collagen gel can be used to replace cartilage and subchondral bone. It has been reported that chondrocytes survived in the fibrin gel and enhanced their synthetic activity as evidenced by the increase of the production of GAG and collagen type II. Human fibrin hydrogels have been approved by the Food and Drug Administration (FDA) for cartilage tissue engineering.

Hyaluronan is a main component of native cartilage. Similarly to the other native biomaterial scaffold, hyaluronan is the most widely used scaffold for cartilage tissue engineering. The studies have shown that hyaluronan upregulated collagen II expression and downregulated collagen I expression in human MSCs when they were cultured in hyaluronan gel. Although bioactive natural scaffolds have very good biocompatibilities, their mechanical properties still need to be improved. In addition to natural bioactive scaffolds, synthetic materials provide good mechanical properties suitable for cartilage tissue engineering. These synthetic polymers are either used alone or combined with natural biomaterials for cartilage research.

The most famous synthetic polymers for cartilage regeneration are polylactic acid (PLA), polyglycolic acid (PGA), and their copolymer polylactic-co-glycolic acid (PLGA). These polymers have a beneficial range of mechanical characteristics and high biocompatibility. Owing to the fact that PLA-PGA polymers have been successfully used in the clinics including sutures, screws, and pins, they are also used for articular cartilage defect repair in rabbits and meniscal lesion repair in dogs. Currently, two PLA-based scaffolds have been clinically used for cartilage repair: one is BioSeed-C and the other one is TRUFIT CB. The PLA-based polymer scaffolds have shown significant improvement in patient outcomes for the treatment of post-traumatic OA and focal degenerative cartilage defects. Polyethylene glycol (PEG), a nontoxic synthetic polymer, is widely used with other natural materials to enhance their mechanical strength for cartilage tissue engineering. The studies have indicated that PEG-based hydrogel can promote chondrogenic differentiation of MSCs *in vitro* and *in vivo*. PEG-HA scaffold-treated patients achieved significantly higher levels of tissue fill in cartilage defects.

New Surgical Techniques for Cartilage Regeneration

Surgical techniques are more important for cartilage repair. In any cartilage repair techniques, the preparation of the defect bed to receive the implant is essential. In order to promote cartilage regeneration, several new surgical techniques have been developed.

Arthroscopic Surgery

Arthroscopic surgery is a common orthopedic procedure that is used in the diagnosis and treatment of problems inside a joint. Generally, the cartilage defect is measured with an arthroscopic graded probe, and the size and the shape of the defect are templated using sterile paper or aluminum that is subsequently used to prepare the graft if it is not an injectable gel form. Besides the defect preparation and measurement, most operations can be done under an arthroscopy. Currently, arthroscopic surgery has been widely used for various damaged cartilage treatments such as degenerative meniscal tear and osteoarthritis of the knee.

Open Surgery

Open surgery is used for some arthroscopically inaccessible cartilage defects such as patella, trochlea, posterior femoral condyle, and some scaffolds that cannot be implanted arthroscopically. This technique has been widely used in cartilage tissue engineering for animal surgery and clinical practice.

Microfracture Surgery

Microfracture surgery is a surgical technique used to repair damaged cartilage by making multiple small holes in the surface of the joint to stimulate a healing response. This technique was developed in the early 1980s by Steadman and his colleagues. The technical details of microfracture have been well summarized. Several animal studies have been completed to assess the microfracture technique. The functional outcomes of patients treated with microfracture for traumatic chondral defects have shown significant improvement. Currently, microfracture surgical technique is considered to be an effective arthroscopic treatment for full-thickness cartilage defect. However, some studies have shown that the younger patients have better clinical outcomes and quality cartilage repair than older patients.

Mosaicplasty Surgery

Mosaicplasty surgery is another common cartilage restoration technique in standard clinical practice. This technique was introduced into clinical application in 1992. Mosaicplasty surgical technique is based on the mosaic-like transplantation of several small, cylindrical plugs of bone and cartilage to provide an even resurfaced area. The long-term clinical follow-up results have shown that the mosaicplasty-treated patients can regain their pre-injury activity level. The studies have demonstrated that the treatment of mosaicplasty in a single cartilage defect size one to five square centimeters of the femoral condyle resulted in clinically relevant better outcome than microfracture.

Cardiac Tissue Engineering

Damage to heart muscle, acute or chronic, has long been considered a tipping point for individual health outlook and progression to heart failure. National statistics are grim, with over 5 million affected by heart failure each year and nearly 300,000 deaths. The problem is that adult heart muscle cells, cardiac myocytes, cannot divide to replace injured cells. Thus, despite a limited population of resident cardiac stem cells, the heart cannot repair itself by any native processes. Instead, scar tissue develops over regions of damaged myocardium. Such scar tissue keeps the organ intact but cannot contract. The ideal clinical intervention would either avoid such scar formation, or simply replace formed scar tissue with functioning cardiac muscle tissue. In a first approach to such therapy, investigators have used injections of new cells into damaged areas of cardiac tissue. These studies have met with limited success due to cell death, exit of cells from the heart, and poor cellular integration with the receiving heart tissue.

It is known that cells have certain requirements for their survival in any surroundings. Many of those relate to the diffusion of available substances, with nutrients and oxygen flowing into cells while waste and signaling molecules are moved out. However, another less well known but essential requirement for nearly all cells is the need for a specific physical substrate. Most cell types, including cardiac myocytes, thrive only when attached to a suitable supporting surface. The purpose of tissue engineering is, then, to create a viable cellular environment through the use of biologically acceptable materials. The idea is that transplantable cells can be contained and organized in so-called engineered scaffolds. Such scaffolds with contained cells can then be used to treat or replace a part of the body. Empty fabricated structures can also be implanted *in vivo*, providing a structure to condition stem cells already present.

For those helped by analogy, the efforts of tissue engineering can be compared to the building of a house. Home construction demands walls for stability, doors for transporting materials, plumbing and circuitry for moving water and electricity, and windows for circulation of oxygen. Analogous matters – anchorage, diffusion, integration, and vascularization, respectively – must be considered when engineering tissue. The purpose of such design is to make a livable structure. In the case of cardiac tissue, healthy, functional cardiac myocytes are the ideal inhabitants.

Identifying Basic Cell Responses

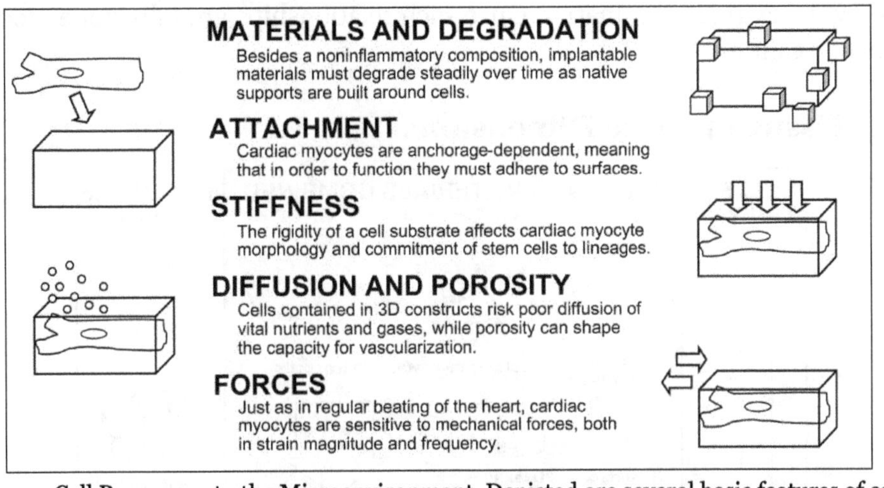

Figure: Cell Responses to the Microenvironment. Depicted are several basic features of cell surroundings that must be taken into account when engineering 3D cardiac tissue.

Flat, two-dimensional (2D) culture dishes have dominated studies of cell biology in the past. However, planar surfaces are hardly fitting for most cell types. To that end, there has been some understanding as to what physical features are important for the design of a three dimensional (3D) structure to support cardiac cells. It is possible, with that understanding, to begin modeling effective local microenvironments for cells.

Slight changes beyond a uniformly flat state can affect cardiac myocytes. Collagen, the most abundant constituent of the extracellular matrix, has a bundle diameter measured in the nanometer range – 100 to 1000 times smaller than the micrometer average cardiac myocyte diameter. Just as cells are influenced by extracellular proteins like collagen, so also are they capable of responding to artificial structures in the same size range. Indeed, cells have been shown to detect surface-deposited fibers manufactured in the nanometer scale. Meshes of electrospun nanofibers support cardiac myocyte adhesion and spontaneous contraction. Projections in the micrometer scale can also directly affect cell behavior. Artificial projections with heights of five micrometers have been shown to cause both changes in cardiac myocyte shape and remodeling of cellular proteins responsible for substrate attachment.

Stimuli other that surface topography can also have robust cellular consequences. Mechanical forces, either generated from cell contraction or sensed from external sources, have pronounced effects on differentiation, growth, and survival of myocardial cells. For instance, experimentally exposing cardiac myocytes to repeated stretching in specific directions alters amounts of contractile proteins. In addition, the stiffness of materials alone can modify some characteristics of cells. Softer, more elastic culture surfaces result in rounded cell shapes, irregular attachments, and higher motility. Furthermore, when material or matrix stiffness is tuned to the range of native muscle rigidity (bearing a Young's modulus – a measure of resistance to deformation – between 8–17 kPa), mesenchymal stem cells are induced to differentiate into a myogenic lineage.

The clinical implications of controlling stiffness can be appreciated when it is recognized that collagen, a primary component of scar tissue that is overproduced in heart failure, has a relatively high stiffness. Though designing around forces and stiffnesses may seem rudimentary, the interplay of these two physical elements has great significance for cardiac tissue engineering. Consider, for example, that each effective beat of the heart constitutes a precise balance between tension of contracting muscle and resistance held by the extracellular matrix. Once such relationships are fully understood, they can be used for therapeutic gain.

Engineering Tissue in Three Dimensions

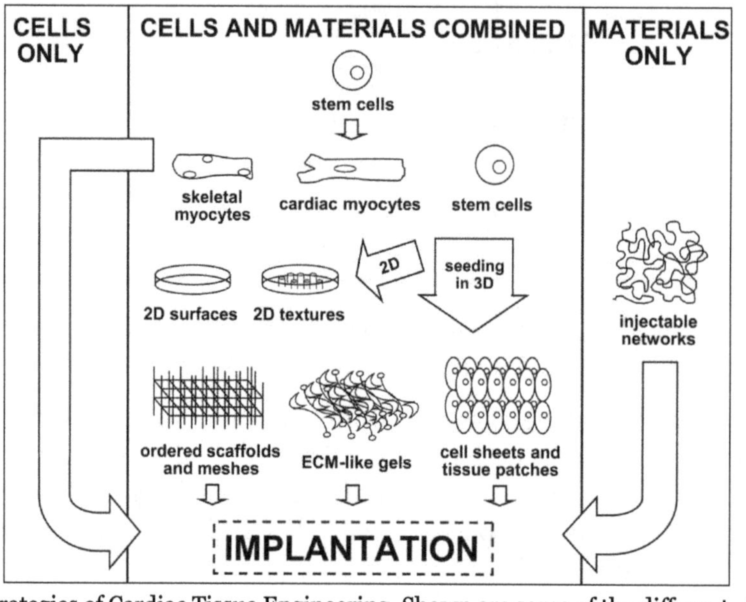

Figure: Strategies of Cardiac Tissue Engineering. Shown are some of the different structural approaches for replacing cells of the myocardium.

The complexity of engineered systems increases considerably with the passage from two to three dimensions. As 2D becomes 3D, a host of new issues becomes significant. Here, levels of diffusion determine the degree of nutrient delivery and metabolic waste removal available to cells in the construct interior. The thickness of normal, diastolic, left ventricular myocardium is a little over one centimeter in the human heart, but there is a 200-micrometer tissue depth limit for the diffusion of oxygen. Therefore, cell survival in 3D cardiac tissue constructs depends on angiogenesis and functional vascular integration to serve the remaining 95% of the heart wall.

Table: Summary of Strategies. Corresponding to the various strategies mapped out in figure, the advantages and disadvantages of notable cardiac tissue engineering approaches are listed.

Method	Clinical Advantages	Clinical Disadvantages
Cells Only		
Injection of cells	No structure to design, less invasive delivery.	High amount of cell exit or loss, poor cell integration.

Cells + Materials		
Ordered scaffolds and meshes	Uniform support for cells, degrade safely as cells incorporate.	Greater surgical risks for implantation, limited vascularization.
ECM–like gels	Similar composition to body, cells readily attach.	Natural components can be inflammatory, slow vascularization.
Cell sheets and tissue patches	High cell density and organization, less obstructive materials.	Surgery and placement difficult, limited vascularization.
Materials Only		
Injectable networks	No immunogenic cells involved, support placed where needed.	May require chemical factors, inadequate repair without cells.
External restraints	No immunogenic cells involved, simple design and function.	Passive device, only prevents spread of damage and stress.

Much of cardiac tissue engineering involves the struggle to perfect cellular organization without sacrificing tissue vascularization. The *in vivo* myocardium has a very dense layout, with overlapping arrays of muscle cells arranged in different circumferential orientations. Individual branched cardiac myocytes in the heart are linked to others at both ends through intercalated discs (containing gap junctions and other adherent sites) that help to transfer both molecular signals (e.g., electrical coupling) and forces of contraction. Studies have verified that a compact, end-to-end cardiac myocyte arrangement enhances electrophysiological connection and function. Replicating this cellular organization of the myocardium has proved difficult.

The simplest 3D constructs that have been investigated to date are tiered 2D structures, in which individual flat levels are overlaid to form a cell-supporting lattice. One such study showed that overlapping electrospun meshes allows attached cardiac myocytes to have strong cellular contacts and synchronous beating after two weeks in culture. In that study, five equal levels were formed with adequate diffusion, each showing *in vivo*-like cell morphology. It is also possible to pattern cells on a single substrate level and builds up into 3D solely by attaching more cells. One group has demonstrated that lanes of cardiac myocytes on degradable polymer films can grow up to three cell layers thick and form organized functional tissue. Such geometric configurations constitute the very definition of tissue – an organized collection of cells working as a unit.

Some investigators have ignored scaffold patterning and studied tissue integrity in irregular environments. Preparations of cardiac myocytes bound to sponge-like alginate, a natural polymer derived from seaweed, have yielded vascularized cylindrical tissue nine weeks after implantation in rats. In another example, porous nonwoven polyglycolic patches were seeded with embryonic stem cells and implanted on the ventricular surface of infarcted hearts in mice. Within eight weeks, the polyglycol had degraded to natural byproducts, leaving a small area of active cells that contributed to an increased survival rate. Indeed, the microenvironmental organization of cell-supporting materials can vary wildly but equally lead to positive experimental effects.

More 3D Approaches and Considerations

Extracellular matrix components such as collagen and fibrin can be used to make elastic gels with compositions similar to the body's extracellular matrix. Gels in unpolymerized form can be mixed with cells, and the result polymerized to create specific geometric shapes. In one study, collagen and cardiac myocytes were combined into circular molds, which displayed interconnected, beating cells when implanted around working rat hearts. Gels and transplantable cells can also polymerize *in vivo* after injection, permitting the cell-matrix composite to assemble and conform to specific areas of the myocardium – though this relinquishes all control over final tissue shape. Two separate studies in rats have used, respectively, skeletal myoblasts in injectable fibrin matrices and embryonic stem cells in collagen matrices. Both studies reported small decreases in heart failure progression. In yet another variation, unpolymerized gel-cell mixtures can be used to fill in the voids of other implantable scaffolds, thereby combining methods and materials into a complex structure for cells. Work of this sort has yielded healthy engraftment in rats. With all these different gel setups, some problems remain. The use of natural proteins can trigger inflammatory responses in the body; the relatively low concentration of cardiac myocytes limits force of contraction; and once again, adequate vascularization can be a challenge – though the addition of endothelial cells (the cell type comprising all vessels) to gel mixtures can hasten new vessel formation.

In what may seem like a total surrender of engineering, another strategy for cardiac regeneration is to simply use the heart's own extracellular architecture. One study described the reseeding of whole decellularized donor rat hearts with major myocardial cell types including cardiac myocytes, fibroblasts, endothelial cells, and smooth muscle cells. According to the report, original cells were removed with a detergent rinse, leaving the entire underlying extracellular matrix intact. In this way valves, chambers, and vascular channels are all preserved, allowing transplanted cells to move in and occupy familiar surroundings. Results were modest – synchronous contractions were observed within several days, generating weak total pump function. Though human use would be feasible with cadaveric hearts (given that the extracellular matrix survives much longer than the cells it supports), the matter of repopulating the structure with appropriate stem cell sources would need to be addressed.

To increase the chance of successful implantation, bioengineers often treat cells in bioreactors before their use *in vivo*. A bioreactor is essentially a container filled with culture media that can be fine-tuned to stimulate cells in a number of ways. Continuous mixing of fluid within bioreactors helps to maintain an even concentration of nutrients. External forces such as pressure, strain, and shear stress can simulate the kinetic features of the heart. In one case, it was shown that a week of incubation in a bioreactor caused cardiac myocytes seeded in polyglycolic acid scaffolds to assemble into uniform, electrophysiological 3D tissue. Another study with cardiac myocyte polyglycolic acid constructs confirmed that tissue thickness could be increased to a few millimeters by perfusing the media over cells. Beyond this, other physical and chemical tweaks to bioreactor environments can enrich the conditioning process but cannot recapitulate the thickness of the ventricular wall in the human heart.

Limiting Cells or Materials

Though they can aid cell attachment and vascularization, the configurations of scaffolds, constructs, and gels are not easily optimized for transferring the contractile forces of the heart. Doing away with implanted materials altogether might have advantages. One group has shown progress in producing intact sheets of cardiac myocytes on a removable substrate, which can then be applied to regions of the myocardium as tissue patches. In the study, three identical layers were carefully stacked and implanted in rat hearts. Since each triple layer had an 80-micrometer thickness, one stack was implanted only to permit blood vessel penetration and cell integration. However, after a short time window, it was possible to repeat the surgery and add a new stack of cardiac myocytes directly over the previous set. By this method, a one-millimeter thick segment of tissue was created after 3 days. Though the stacks of cardiac myocytes were active and synchronized, the repetitive surgery technique has obvious difficulties for human treatment. Another group tested this same idea in rats with sheets of mesenchymal stem cells, which caused moderate recovery of cardiac function through vessel formation and cardiac myocyte differentiation. These cell sheets may owe their effectiveness to the fact that cardiac myocytes are able to drive their own organization into rhythmically-beating aggregates – without cues from a matrix or substrate.

If all the technical problems of bioengineering functional heart tissue in animal models were solved, there would still be a major problem: there is at present no practical source of donor cells for replacement of human myocardium. Studies involving committed cardiac myocytes in scaffolds or constructs are almost universally taken from neonatal or adult rat hearts. Though such animal models have a physiological likeness that is well-suited for research, they represent a supply that is unacceptable for clinical use. Adult stem cells present the only feasible source for human treatment, but isolation, expansion, and differentiation into contractile cells remains difficult. Cell populations such as embryonic stem cells, mesenchymal stem cells from bone marrow, and umbilical cord stem cells have shown some therapeutic promise, though non-autologous cells bear an added risk of immune rejection (as detailed in both preceding articles in this series). Very large quantities of differentiated cardiac myocytes are the necessary endpoint; a human myocardial infarction can injure 50 grams of cardiac muscle tissue, which would call for substitution of over one billion cells. Any success in building the supply of cardiac myocytes from stem cells could eventually be coupled to the discussed housing concepts of tissue engineering.

Tissue engineering approaches need not all be cell-based. Pharmacologic agents can be linked to implanted scaffolds for controlled release. One of the more advanced cardiac-related examples of this makes use of injectable peptides that self-assemble into randomly branched networks; the resulting nanofiber microenvironments can be used to deliver drugs or growth factors. Remarkably, an implanted structure alone can aid survival of myocardial cells following injury. For instance, fibrin matrices have been shown to increase vascularization, reduce scar formation, and preserve cardiac function when injected to damaged regions in rat hearts. This concept can be expanded further, as demonstrated by the availability of external restraints. These bare polymeric supports are intended to wrap around the outer surface of one or both ventricles, thereby preventing wall stresses and maladaptive shape changes that may precede heart failure. Indeed, in a conceptually simple application, clinical trials of degradable, passive restraints placed around the heart showed enhanced total cardiac function in hundreds of patients. Though such improvements may be modest, this method of design represents a way of thinking that may produce new breakthroughs. Circumstances that would actually benefit from minimized engineering should not be overlooked.

Bone Tissue Engineering

The Need for Engineered Bone

While bone is inherently capable of regeneration, complications such as excessive bone loss impede healing, necessitating the use of bone grafts. In the United States alone, an estimated 15 million fractures occurs annually, including 1.6 million hospital admissions for traumatic fractures and 2 million osteoporotic fractures, costing over 60 billion dollars and calling for 1.6 million bone grafts each year; a growing demand for bone grafts is similarly observed worldwide. In such applications, autologous bone grafts continue to be regarded as the “gold standard” for bone repair. However, this may not be practicable in cases involving large bone loss. Additionally, patients suffer from significant donor site morbidity, as well as poor outcomes in older patients. Allogeneic bone grafts may alternatively be used, but pose potential risks of immune rejection and pathogen transmission. Additionally, the limited number of donors is unable to cope with the clinical demand. Consequently, alternative approaches to provide efficacious and reliable bone grafts are being actively pursued.

The advent of tissue engineering, where the aim is to generate functional tissue, has raised the possibility of engineering bone in vitro. Over the past few decades, a wealth of progress in bone tissue engineering has been achieved, particularly in cell sources, developing biocompatible and biodegradable scaffolds, designing bioreactors to enhance in vitro osteogenic priming, and identifying growth factors that can induce or promote endogenous bone and vascular formation. Numerous pre-clinical trials with various animal models have produced optimistic results. Despite the initial optimism, the lack of translation into a clinical setting suggests significant issues remain, including optimisation of cell sources, choice of biomaterial, in vitro preparation and the mode of delivery.

Normal Bone Anatomy

As seen in figure, bone tissue is organised into cancellous or cortical bone. Cancellous bone (also referred to as trabecular bone or spongy bone), is porous, providing structural support and organisation for bone marrow interspersed inside. In contrast, cortical bone is the compact bone surrounding the marrow space, and confers mechanical strength to bone. The outer layer that covers the surface of cortical bones is the periosteum, containing mainly blood vessels and osteoblasts, which are activated during appositional growth and bone repair.

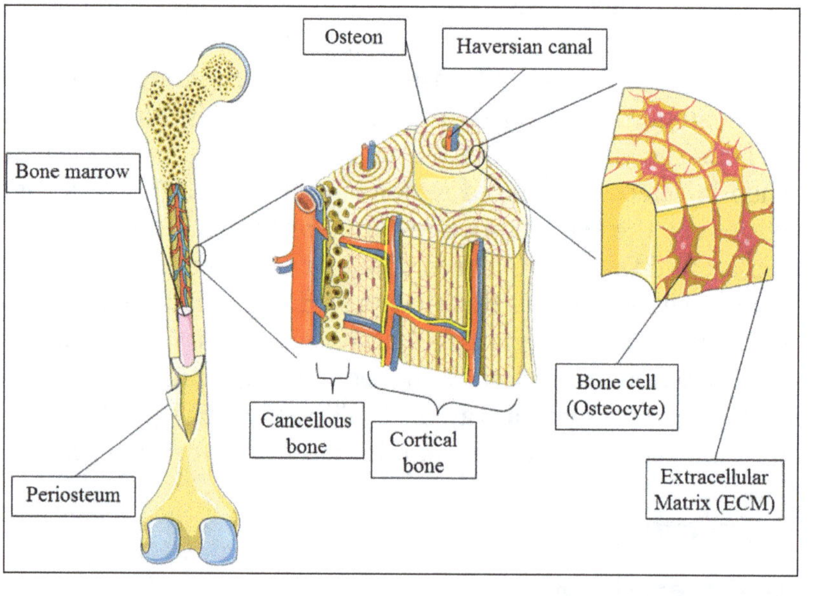

Figure shows schematic overview of bone, depicting gross overview, and cellular distribution. (Figures were produced using Servier Medical Art) In particular, osteoprogenitors may be found abundantly in the periosteum and bone marrow, where they perform critical roles in bone repair. Additionally, bone is observed to be highly vascularized, in both the intramedullary canal and periosteal region.

The basic functional unit of the cortical bone is the osteon (or haversian system) which contains cells and extracellular matrix organised in a lamellar pattern, surrounding the haversian canal, in which nerves and blood vessels reside. Within the osteon, osteoblasts and the mature osteocytes exist, contributing to the generation and maintenance of ECM that gives bone it's structural strength. These are derived from osteoprogenitors that reside in the bone marrow and periosteum. Also present are the myeloid-derived osteoclasts that mediate bone resorption; further elaboration on the roles and functions of these cells is provided below. Compared to other tissue, the ECM found in bone is highly mineralized to confer mechanical strength. Calcium and phosphate are found in bone ECM in the form of hydroxyapatite crystal ($Ca_{10}(PO_4)_6(OH)_2$), interspersed with Type I collagen, while the exact composition remains to be elucidated. This unique composite of biological molecules (collagen) and inorganic minerals (calcium phosphate) provides bone with high mechanical strength, as well as toughness to resist impact.

Bone Physiology

Bone is a highly dynamic tissue, which undergoes constant remodelling through one's lifetime. Homeostasis is achieved through the combined actions of osteoprogenitors, osteoblasts, osteocytes, as well as osteoclasts. In general, bone formation is effected through the proliferation of osteoprogenitor cells and differentiation into osteoblasts, which are responsible for the regulation of mineralization and collagen production through expression of functional proteins such as osteocalcin and alkaline phosphatase. Osteoblasts which get trapped in the ECM eventually differentiate to mature osteocytes forming syncytial networks that function to support bone structure and metabolism. Bone actively undergoes remodelling in response to environmental stimuli, such as physiologic influences or mechanical forces, in accordance with Wolff's law. This process involves a balance between bone formation as described above, as well as bone resorption by osteoclasts, and it is crucial for renewing bone, to maintain bone strength and mineral homeostasis.

The bone is able to undergo significant regeneration in response to trauma and fractures. Regeneration progresses through the three phases of inflammation, repair and remodelling. The immune system can be seen to be intricately involved in the process of healing and inflammation is required for effective healing. Subsequently, the repair and remodelling processes are initiated in the periosteum, which contain a rich population of osteoprogenitors and osteocytes that proliferate and migrate to the defect site, forming a bony bridge to effect fracture healing. When this bridging does not happen, hypertrophic non-union occurs, necessitating the use of bone grafts.

Stem Cell Sources

Over the last few decades, stem cells have emerged as a key player in tissue engineering, both for in vitro generation of bones, and in vivo bone regeneration. Two hypotheses have been proposed to describe the mechanism of healing affected by stem cells. In the conventional approach, osteogenic stem/progenitor cells are proposed to participate in new bone through direct differentiation into functional tissue. More recently, it has emerged that trophic factors secreted by administered stem/progenitor cells can promote functional tissue regeneration. Much research is currently focused on understanding the influence of stem cell origin and culture conditions on clinical outcome.

Adult Stem Cell Sources

The human body retains a small amount of osteoprogenitor cell population with differentiation potential even in adulthood. The main osteogenic progenitor cells – mesenchymal stem cells (MSCs) are considered to consist of a variety of cell populations, with no unique marker to delineate a MSC to-date. Therefore, in some studies, the term "mesenchymal stem cells" is used to represent osteoprogenitor cells. The current consensus on the characteristics of a bona-fide MSC should include: 1) expression of cell-surface markers which are non-haemopoietic, non-endothelial and incorporates a number of other cell-surface markers, including CD73, CD105 and CD90, and 2) the capacity for tri-lineage differentiation into osteocytes, chondrocytes and adipocytes. Common sources of MSCs include bone marrow, fat tissue and the periosteum.

MSCs are most abundantly found in the bone marrow (BM), and BM-MSCs are widely utilised for either autologous or allogeneic transplantations. BM-MSCs can be easily expanded in culture, and induced to differentiate into osteoblasts through culture in a high phosphate environment. In tissue engineering applications, BMMSCs have been used to generate bone graft by loading into appropriate scaffolds, followed by osteogenic culture,. BM-MSCs may offer favourable characteristics, including availability as an autologous cell source, and thus, non-immunogenic cell source. In contrast to suggestions that BM-MSC are immunoprivilleged and thus suitable for allogeneic – unmatched transplantation, a few studies have emerged that reported rejection of murine BM-MSCs following infusion into immune competent recipients. Besides, adult BM-MSCs though advantageous in autologous source availability, have shown relatively low efficiency in osteogenic differentiation in vivo in several comparative studies.

The periosteum is involved in extraosseous fracture repair, and is thus also thought to contain osteoprogenitors. Indeed, stem cells derived from periosteal tissue showed better mineralization and neovascularization ability than adult bone marrow derived cells and demonstrating good efficacy in healing a calvarial defect model. However, its potential as a cell source is limited by the low volume of periosteum which can be harvested, as well as the complexity of the harvesting process. Further optimization in terms of isolation and culture conditions will be required before it can be a practical source.

Adipose tissue represents another potential source of osteogenic stem cells. Adipose derived stem cells (ADSCs) can be isolated from fat tissues harvested by liposuction or surgical fat section, which are typically considered medical waste. Similar to BM-MSCs, ADSCs respond to osteogenic induction through upregulation of alkaline phosphatase and other key osteogenic proteins, and differentiate into osteoblasts capable of depositing minerals. Additionally, ADSCs have been shown to ne efficacious in bone repair, facilitating the repair of critical size defects in both calvarial and femoral segmental defects. However, the efficacy of bone repair by using ADSCs alone is relatively low compared with marrow-origin MSCs and more primitive MSCs. The adult stem cells represent a rich autologous source for bone tissue engineering, and the long-history study accumulate mature methodology to facilitate their application in pre-clinical trials or even clinical trial. Yet the low efficiency of in vitro bone generation and in vivo bone regenerative capacity may call for optimized induction strategy or alternative sources.

Embryonic Stem Cell (ESC) and Induced Pluripotent Stem Cell (iPSC)

In the last decade, the osteogenic induction of embryonic stem cells (ESCs) and the creation of induced pluripotent stem cells (iPSCs) presented new cell sources for bone tissue engineering. However, expansion of ESCs and iPSCs to clinically useful numbers is logistically challenging, and autologous use is not possible in the case of embryonic stem cells. Moreover, precise control of differentiation is necessary before applying to clinical use. This is especially so since the presence of undifferentiated "rogue cells" may result in tumour formation after transplantation.

ESCs are defined by their pluripotency, and can be induced into osteogenic cells through different methods in vitro. It is showed that human ESCs can be induced into osteogenic cells and form mineralized bone in vivo without presence of teratoma. However, more investigations will be required to address the potential of teratoma in embryonic stem cells transplantation. Mature methods for specific osteogenic differentiation are also needed to overcome non-specific differentiation before clinical implementation can be contemplated.

As an alternative, iPSCs have recently emerged as a potential cell source for regenerative medicine. Compared to ESCs, iPSCs face fewer ethical challenges, and are able to serve as an autologous cell source. iPSCs have been successfully differentiated into osteoblasts like cells, or through induction into a MSC phenotype, allowing ready expansion of the iPSC-derived MSC before their direct differentiation into osteoblasts. However, similar issues to that of ESC remain, and concerns remain over the potency of the cells. Additionally, the concept of reprogramming remains poorly understood and leaving questions on the fate of cells in vivo. Therefore, although the differentiation capacity qualifies the iPSC as a potential source for bone engineering, yet a lot more efforts have to be done before it can be competent.

Fetal Stem Cell Sources

Compared with adult stem cells, fetal stem cells are relatively more primitive, and have higher proliferative and differentiation capacity. MSCs have been derived from various fetal tissues in early to mid-gestation, including the blood, bone marrow, liver, pancreas, kidney thymus and bone. Of these, fetal BM-MSC has shown particular utility for bone tissue engineering. However, its availability may be limited by ethical issues especially where fetal tissues are used to derive MSC.

As an alternative source, MSC has also been derived from umbilical cord blood at term gestation, albeit at very low frequencies, with optimised protocols achieving up to 60% success in its derivation. Aside from the cord blood, MSC has also been derived from the umbilical cord vessels and matrix, the placenta and fetal membranes. Although these sources of MSC are plentiful and readily harvested, their utility for bone tissue engineering is somewhat limited compared to fetal bone marrow derived MSC.

Neural Tissue Engineering

Human brain is just like black box of an airplane. It has so many processing centers which are required for proper functioning of life activities. Unit of human brain is nerve cell. So for the treatment of any neural disorder or neural damage we will have to concentrate on the repairing of nerve cell. There is a technique called Autologous Nerve Graft for repairing of neural damage. This techniques is also called –gold standard for repairing nerve damage. This is not a perfect treatment because the final outcome is only limited function recovery. It also requires multiple surgeries to harvest the tissue and to implant it. Other treatment strategies include Allograft and Xenografts.

In allograft, the tissue for graft is taken from another person while in xenograft; tissues are taken from any other species. Both allografts and xenografts have same disadvantages as autografts, but additional problem here is tissue rejection from immune response.

These disadvantages can be solved with the help of Tissue Engineering (TE). TE basically makes use of a combination of cells, growth factors and scaffolds to improve or replace injured tissues. Scaffold is one of the most important elements of TE. Scaffold, a 3D porous architecture consisting of various biomaterials, acts as a temporary platform for the growth of cells. An ideal scaffold should be biocompatible, less inflammatory, and doesn't produce toxic degradation products. It must be porous for vascularization and cell migration. Scaffolds are in general made of various types of synthetic and natural polymers, such as PGA (Poly(glycolic acid)), PLA (poly (lactic acid), PLGA (Poly (lactic acid-co-glycolic acid)), PCL (poly(caprolactone)), PHB (poly hydroxybutyrate), collagen, gelatin, alginate, silk etc. There are various methods for scaffold fabrication e.g. electro spinning, phase separation, freeze drying, self-assembly, etc.

How Nerve Cells Repair Themselves?

Nerve cells are able to repair the damage of 6 mm. If size of nerve damage is more than that of 6 mm size than it become difficult to repair nerve damage and it is still a challenge. Figure is showing the various events which took place during nerve regeneration. After division of the same single nerve fiber, Wallerian degeneration occurs. The single nerve fibers will –sprout to form a regenerating unit. As the regenerating unit matures, the individual sprouts and become myelinated. Basal lamina adjacent to this regenerating unit represents an area of degeneration without regeneration.

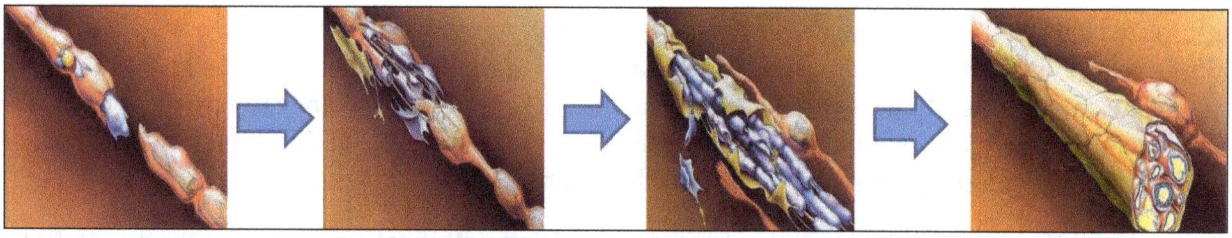

Figure: Various events taking place after nerve degeneration.

If we can enhance the activity of Band of Bungner, then it will ease the process of nerve repairing. Many strategies have been attempted to improve the regenerative potency of neurons such as cell therapy, exogenous delivery of growth factors and tissue engineering approaches; each in turn restores the function with varying degrees of success. Schwann cells have capability of neural regeneration and remyelination. They perform this function by secreting adhesion molecules L1 and neural cell adhesion molecule (N-CAM), extracellular molecules (collagen and laminin) and a number of tropic factors such as nerve growth factor (NGF), brain-derived neurotrophic factor (BDNF) and neurotrophin-3 (NT-3).

Scaffolds and their Possible Role in Neural Cell Repairing

The differentiation of neural stem cells can be controlled by constructing scaffolds of composite biomaterials that have similar properties as that of extracellular matrix (ECM). Stem cells have capability to differentiate in any type of cell; and these stem cells can be proliferated and differentiated in scaffold made up of fibrin like structure. Astrocytes help in development of neural cell or neuro-genesis from stem cell. So we can grow astrocytes on polymeric substance for neuro-genesis. OEC (Olfactory Ensheathing Cells) also have similar properties like that of Astrocytes cell, so OEC can also be used for neural regeneration.

Growing stem cell on polymeric surface seems to be easier but actually it is not so. In this process it is very difficult to control cell proliferation as well as to maintain 3D structure of cells. If we provide various bioactive factors such as NGFs (Neural Growth Factors) to stem cells used in neural tissue engineering then growth of implanted cells on polymer can be increased. These NGFs can be made to reach to target stem cell by making use of guidance channels or conduits. Various unanswered questions in stem cell-based therapy such as carcinogenicity and ethical concerns helps in limiting their use. Electrospun nanofibrous scaffold have become a successful, popular and safe vehicle for delivery of bioactive growth factors and it can simultaneously act as a scaffold also.

Materials for Scaffold Fabrication

Various natural polymers (e.g. chitosan, chitin, collagen, gelatin, alginate), synthetic polymers which are nonbiodegradable (e.g. silicone), synthetic biodegradable polymers (e.g. PLGA, poly (ε-caprolactone (PCL), poly Llactic acid (PLLA)) and conducting polymers (e.g. polypyrrole, polyaniline) have been used for scaffold fabrication for neural tissue engineering. Whatever material is used for scaffold fabrication but in the end fabricated scaffold must be thin, biocompatible, porous, biodegradable, biocompatible and inductive in nature; otherwise fabricated scaffold will be of no use. It should also have appropriate mechanical strength as well as flexibility. It should be pliable and harmless to surrounding tissues. It should also be capable of avoiding structural collapse during implantation. Surface erosion of scaffold is also required for ensuring its long life, because if bulk erosion is taking place then life of scaffold will be small.

How to Increase Cell Adhesion Properties of Scaffold?

Three dimensional tissue culture is important because it increases cell-cell interaction and which in turn improves regeneration of the injured tissue. All these will happen only and only if cells are in proper attachment with scaffold. Proper attachment of cells with scaffold requires good cell adhesion properties of scaffold. Cell adhesion property is a must requirement for any scaffold because if cell will not adhere to scaffold then cell proliferation and differentiation will be a tedious task. Cell adhesion property of a scaffold mainly depends on its surface characteristics such as charge density and wettability. Most of the polymers used in scaffold fabrication are hydrophobic in nature and that's why their adhesive properties are poor. We can improve their adhesion properties by using ECM (Extracellular Matrix) materials such as laminin, fibronectin, collagen etc. All these are basically proteins and when they are coated on surface of scaffold then adhesive properties of scaffold is improved which in turn increases the cell adhesion property. Beside these we can also incorporate specific adhesion peptide sequences like RGD and IKVAV, YIGSR. They induce hydrophilicity which in turn improves the cell adhesion property of the scaffold.

What are the Available Techniques for Scaffold Fabrication?

There are different scaffold fabrication techniques such as solvent casting, particulate leaching, melt molding etc., to fabricate scaffolds of various geometries with the desired porosity and surface area for cell scaffolding. Scaffolds made up of nanofibers, sintered matrix, nanofoams, hydrogels, nanotubes, etc., have been extensively used as a scaffold in neural regeneration.

S. No.	Biomaterials	Modification/Method of fabrication	Improved Properties
1	Star-Poly(ethylene glycol)	Incorporation of polysaccharide (Heparin).	Tunable physical and mechanical properties to adopt specific tissue requirements.
2	Chitosan	Modified with (γ- glycidoxypropyltrimethoxys ilane (GPTMS).	Mechanical strength.
3	Poly(sialic acid)	Hydrogel modified with adsorbed poly-L-lysine or poly-L-ornithine or laminin or collagen.	Mechano compatibility; Cell adhesive property.
4	Poly(β-hydroxybutyrate)	Sheets impregnated with extracellular matrix Molecules.	Cell adhesion and proliferation.
5	Poly(ε-caprolactone)	Electrospinning and Thermal fiber bonding.	Mechanical strength.
6	Poly(ε-caprolactone)	Aligned fibers by Electrospinning.	Contact guidance.
7	Poly(lactic-coglycolic acid)	Modified immersion precipitation method.	Selective permeability; Hydrophilicity.
8	Poly(D, L-lactideco-εcaprolactone) [PDLLA/CL]	PPy coating substrate and PPy nanoparticle/PDLLA/CL composite.	Electrical cue for multitude of cell functions.
9	Chitosan	Polylysine-functionalised thermoresponsive chitosan hydrogel.	Injectable scaffold; Mechano compatibility; Surface property (wettability, charge density).
10	Poly(ε-caprolactone)	Electrospinning (Polymer blending with collagen).	Biological property (schwann cell adhesion, migration and differentiation).
11	Collagen	Hydrogel crosslinked with YIGSR peptide modified Dendrimers.	Biological function (promote the growth of corneal epithelial cells and neurite outgrowth).
12	Poly(glycerolsebacate)	Replica molding.	Micropatterned substrates; Flexibility; Surface degradable; Strong contact guidance response.
13	Poly(lactic-coglycolic acid)	Microbraiding method.	Flexibility; Porosity.
14	Poly(D,L-lactide-coglycolide)	Low pressure injection molding.	Porosity; Longitudinally aligned channels; Mimics the geometry of native nerves.

Various approaches have been attempted to improve the properties of common biomaterials to make scaffold suitable for neural tissue engineering. Table is showing a list of biomaterials, their method of fabrication and their improved properties. All these tabulated materials have been used in neural tissue engineering.

Electrospinning Technology for Nanofibres Fabrication

Substrate is the essential requirement forfor tissue engineering because it provides an attachment site for cells. Without attachment site the scaffold will be of no use. Nano fibres resemples ECM (Extra Cellular Matrix) and that's why they provide a suitable site for attachment to cell. Diameter of nanofibres is less than 1 micron. Nano fibres have high surface are to volume ratio and they also have very high mechanical strength. Both of these properties are must require for neural tissue engineering. Electro-spinning is the easiest method to fabricate nanofibres. Aligned nanofibres support proper orientation of cell in scaffold and they also improve neurite outgrowth with proper differentiation of neural cell. The parameters such as viscosity, conductivity, surface tension of polymer solution, applied electric potential, flow rate, and distance between the electrodes are to be optimized while carrying on the electrospinning process.

Role of Electrical Stimulation in Neural Tissue Engineering

Neural cells communicate through generating action potential at their synapse. Generating action potential is a type of electrical stimulation. So we can say that electrical stimulation plays an important role in neural communication. So a scaffold which is going to be used for neural tissue engineering must have conducting polymer so that electrical stimulation can easily be pass through scaffold, which in turn will be responsible for better neurite outgrowth. Polypyrrole (Ppy), is a conducting polymer and it is used in biomedical applications. It has capability to enhance the nerve regeneration by electrical stimulation. Besides showing conductivity, Polypyrrole does not produce any type of anti-oxidant, so it is biocompatible, which is its further advantage. Both of these properties make it a suitable material for scaffold fabrication for neural tissue engineering. Cell adhesion property of polypyrrole can be improved by chemical conjugation of carboxylic acid group with RGD peptide.

Future of Neural Tissue Engineering

Neural Tissue Engineering is providing a promising method for neural regeneration, but this area is still not fully explored. Long term effects of usage of polymers on human heath are still to be finding out. Treatment of neurological disorders with the help of neural tissue engineering is still to be done.

References

- Central-dogma-inheritance-mechanism, biology: byjus.com, Retrieved 03, Jan 2020
- What-is-gene-expression, science-education-10797: jove.com, Retrieved 23, Feb 2020
- Gene-cloning-requirements-principle-steps-applications: microbenotes.com, Retrieved 25, june 2020
- Cloning-and-genetic-engineering, biology: opentextbc.ca, Retrieved 07, March 2020
- Gene-cloning-requirements-principle-steps-applications: microbenotes.com, Retrieved 25, April 2020

Permissions

All chapters in this book are published with permission under the Creative Commons Attribution Share Alike License or equivalent. Every chapter published in this book has been scrutinized by our experts. Their significance has been extensively debated. The topics covered herein carry significant information for a comprehensive understanding. They may even be implemented as practical applications or may be referred to as a beginning point for further studies.

We would like to thank the editorial team for lending their expertise to make the book truly unique. They have played a crucial role in the development of this book. Without their invaluable contributions this book wouldn't have been possible. They have made vital efforts to compile up to date information on the varied aspects of this subject to make this book a valuable addition to the collection of many professionals and students.

This book was conceptualized with the vision of imparting up-to-date and integrated information in this field. To ensure the same, a matchless editorial board was set up. Every individual on the board went through rigorous rounds of assessment to prove their worth. After which they invested a large part of their time researching and compiling the most relevant data for our readers.

The editorial board has been involved in producing this book since its inception. They have spent rigorous hours researching and exploring the diverse topics which have resulted in the successful publishing of this book. They have passed on their knowledge of decades through this book. To expedite this challenging task, the publisher supported the team at every step. A small team of assistant editors was also appointed to further simplify the editing procedure and attain best results for the readers.

Apart from the editorial board, the designing team has also invested a significant amount of their time in understanding the subject and creating the most relevant covers. They scrutinized every image to scout for the most suitable representation of the subject and create an appropriate cover for the book.

The publishing team has been an ardent support to the editorial, designing and production team. Their endless efforts to recruit the best for this project, has resulted in the accomplishment of this book. They are a veteran in the field of academics and their pool of knowledge is as vast as their experience in printing. Their expertise and guidance has proved useful at every step. Their uncompromising quality standards have made this book an exceptional effort. Their encouragement from time to time has been an inspiration for everyone.

The publisher and the editorial board hope that this book will prove to be a valuable piece of knowledge for students, practitioners and scholars across the globe.

Index

www.ingramcontent.com/pod-product-compliance
Lightning Source LLC
LaVergne TN
LVHW080736210926
848485LV00003B/567